SpringerBriefs in Applied Sciences and Technology

T0207244

For further volumes:
http://www.springer.com/series/8884

Dario Croccolo · Massimiliano De Agostinis

Motorbike Suspensions

Modern Design and Optimisation

 Springer

Dario Croccolo
Massimiliano De Agostinis
DIEM
Universita Di Bologna
Bologna
Italy

ISSN 2191-530X   ISSN 2191-5318   (electronic)
ISBN 978-1-4471-5148-7   ISBN 978-1-4471-5149-4   (eBook)
DOI 10.1007/978-1-4471-5149-4
Springer London Heidelberg New York Dordrecht

Library of Congress Control Number: 2013936525

Printed on acid-free paper

Springer is part of Springer Science+Business Media (www.springer.com)

Contents

Acronyms

Ar_a	Rolling friction force applied on the front wheel
Ar_p	Rolling friction force applied on the rear wheel
f_a	Rolling coefficient of friction on the front wheel
Fa	Aerodynamic force
F_b	Braking force
F_{chain}	Chain force
Fc	Total centrifugal force
Fc_a	Centrifugal reaction applied on the front wheel
Fc_p	Centrifugal reaction applied on the rear wheel
F_d	Braking disk force
Fi	Inertia force
f_p	Rolling coefficient of friction on the rear wheel
Fp	Total load generated by masses
Fp_a	Vertical reaction applied on the front wheel
Fp_p	Vertical reaction applied on the rear wheel
F_{pin}	Braking pin force
F_t	Traction force
g	Gravity acceleration
m	Total mass
r	Curve radius
v	Motorbike speed
ϕ_c	Chain crown diameter
ϕ_p	Chain pinion diameter
W	Traction power
ω	Rear wheel speed

Chapter 1
Motorbike Equilibrium

Abstract This chapter aims at defining the equations that govern the motorbike equilibrium and are useful for designing the front and the rear suspensions. Such equations are, in some case, non-linear and implicit so that they can be solved by recursive methods applied and presented in the Chap. 3 of this book. The mathematical models can be applied to different types of motorbike but they are developed for the peculiar case of those that adopt a single braking disk installed on front suspension and with a unique arm for the rear suspension.

1.1 Dynamic Equilibrium of the Motorbike

The dynamic equilibrium of a motorbike riding into a straight plane or in a bend, has to be referred to the initial static condition represented in the scheme of Fig. 1.1.

The dynamic equilibrium depends on the loads system applied to the motorbike which can be represented by the scheme of Fig. 1.2. Based on this scheme it is possible to write the following six basic formulae collected into Eq. 1.1.

$$-Ar_p + F_t - Fb - (Fa + Fi) - Ar_a = 0$$
$$Fp_p - Fp + Fp_a = 0$$
$$-Fc_p + Fc - Fc_a = 0$$
$$-Fp \times Y_G \times \sin\gamma + Fc \times Y_G \times \cos\gamma = 0 \quad (1.1)$$
$$(Fa + Fi) \times Y_G \times \sin\gamma - Fc \times X_G rear + Fc_a \times (X_G front + X_G rear) = 0$$
$$(Fa + Fi) \times Y_G \times \cos\gamma - Fp \times X_G rear + Fp_a \times (X_G front + X_G rear) = 0$$

Given that $Fp = m \times g$, $Fc = m \times a_c = m \times \frac{v^2}{r}$, $Ar_a = Fp_a \times f_a$ and $Ar_p = Fp_p \times f_p$, the previous formulae can be combined in order to define the other forces acting on the wheels. These formulae are collected into Eq. 1.2.

$$Fp_a = \frac{Fp \times (X_G rear + f_p \times Y_G \times \cos\gamma) - Y_G \times \cos\gamma \times (Ft - Fb)}{X_G front + X_G rear - Y_G \times \cos\gamma \times (f_a - f_p)}$$

D. Croccolo and M. De Agostinis, *Motorbike Suspensions*,
SpringerBriefs in Applied Sciences and Technology,
DOI: 10.1007/978-1-4471-5149-4_1, © The Author(s) 2013

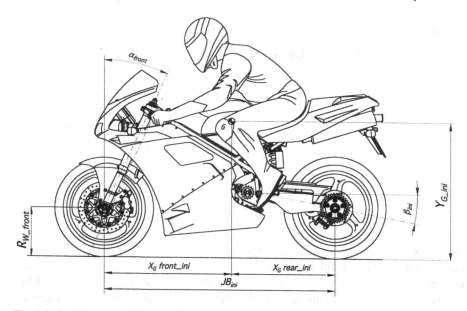

Fig. 1.1 Initial static equilibrium of a motorbike

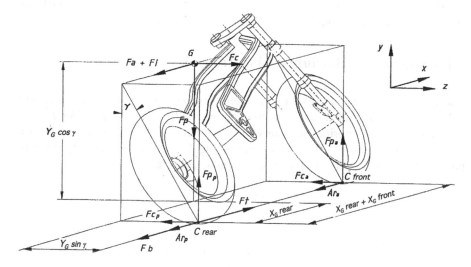

Fig. 1.2 Scheme of forces and reactions acting on the motorbike

$$Fp_p = \frac{Fp \times (X_G front - f_a \times Y_G \times \cos\gamma) + Y_G \times \cos\gamma \times (Ft - Fb)}{X_G front + X_G rear - Y_G \times \cos\gamma \times (f_a - f_p)}$$

$$Fc_a = Fp_a \times \tan\gamma \quad (1.2)$$

$$Fc_p = Fp_p \times \tan\gamma$$

$$\gamma = arctan\frac{v^2}{r \times g}$$

The previous formulae depend on:

- Masses of vehicle, pilot and passenger.
- External forces.
- Center of gravity position.
- Roll angle.
- Motorbike speed.
- Curve radius.
- Steering angle.

The masses can be considered as constant whereas the other quantities are variable. For instance the curve radius depends on the steering angle, while the roll angle depends on the motorbike speed, on the curve radius, on the gravitational acceleration and, obviously, on the grip conditions between the tyre and the ground. Some further considerations concerning the grip will be presented in the next chapter. Once the grip limit and the masses distribution are known, the equilibrium equations can be directly solved.

Unfortunately the center of gravity depends on:

- Roll angle.
- Steering angle.
- Driver style.
- Diving and rising of suspensions.

Diving and rising of suspensions depend on the masses distribution and on the position of the center of gravity. Thus the previous equations are implicit and can be solved, via numerical methods, starting from an initial position of the center of gravity (for instance the static initial position represented in Fig. 1.1), which has to be corrected up to the equilibrium convergence of the formulae of Eq. 1.2. Furthermore the rear suspension could have a progressive stiffness and a geometric configuration that makes the convergence even more difficult.

Since the system of equations holds true for every different driving condition, the use of a calc sheet provided with a solver engine, is the best way of calculating the parameters and of highlighting the worst load conditions.

In the following sections the solution procedure will be presented and described giving, also, some comments concerning the use of variables.

1.2 Steering Angle

The steering angle produces the diving of center of gravity as indicated in the scheme of Fig. 1.3 [1, 2]. The value of the displacement Δh for tubular wheels with wheel offset d equal to zero (see the sketch of Fig. 1.4) can be calculated by Eq. 1.3 in which R_W is the wheel radius, Δ is the steering angle and ε the caster angle.

Anyway, for a large number of motorbikes, the variation of the center of gravity due to the steering angle and to the curve radius, is negligible since it results within

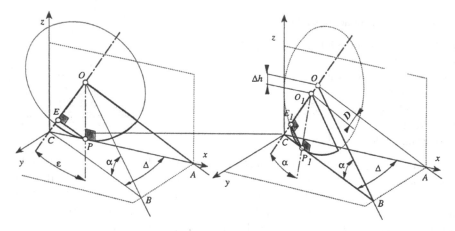

Fig. 1.3 Diving of center of gravity due to the steering

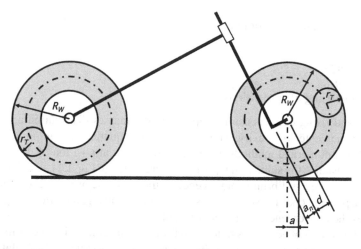

Fig. 1.4 Motorbike geometric parameter

the range from −1 to 1 %. Therefore the steering angle and the curve radius do not influence the roll angle α indicated in Fig. 1.5 which depends only on the friction limit between the wheels and the ground and on the motorbike speed. The present hypothesis makes the computational process easier even if the result values can be accepted when the motorbike speed and the steering angle are compatible with the roll angle.

$$\Delta h = \left[1 - \sqrt{1 - \sin^2 \Delta \times \sin^2 \varepsilon}\right] \times R_W \tag{1.3}$$

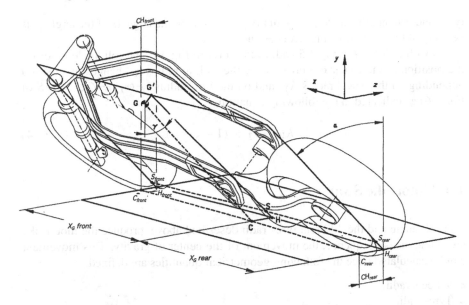

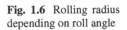

Fig. 1.5 Driving style influence

Fig. 1.6 Rolling radius
depending on roll angle

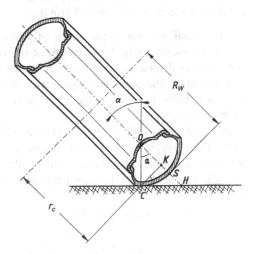

1.3 Driving Style and Roll Angle

The driving style modifies the center of gravity position. Usually the center of gravity
location moves towards the center of the curve as indicated in the scheme of Fig. 1.5:
the γ angle increases while the roll angle α remains the same.

Since the tyres have a circular section and γ is the angle between the vertical plane
and the plane containing the center of gravity and the two contact points between

tyres and ground, the angle γ is, actually, very close to the angle α; if the angle γ is kept equal to the angle α, the errors are negligible.

As well indicated in Fig. 1.5 and, even more, in Fig. 1.6, the roll angle modifies the position of the center of gravity since the rolling radius r_c changes its position depending on the wheel radius R_W and on the tyre radius r_T ($r_T = \overline{OC} = \overline{OS}$ of Fig. 1.6) as indicated in the following equation:

$$r_c = R_W - r_T \times (1 - \cos\alpha) \tag{1.4}$$

1.4 Motorbike Sway

The roll angle and the different tyres radii described above, produce the motorbike sway and, as consequence, the movement of the center of gravity. This movement can be calculated once the following geometrical quantities are defined.

- Wheels radii.
- Tyres radii.
- Initial distance between the wheel axles.
- Initial position of the center of gravity.

The actual position of the center of gravity can be defined as the sum of two contributions: the sway displacement of the front axle keeping the rear axle fixed and the sway displacement of the rear axle keeping the front axle fixed.

Referring to the geometrical quantities indicated in the sketches of Figs. 1.7 and 1.8 it is possible to calculate the actual position of the center of gravity depending on the roll angle α.

Firstly, the sway of the front wheel is calculated keeping the axle of the rear wheel fixed. The geometric relationships are indicated in Fig. 1.7 in which G_0, J and B are the center of masses, the axle center of the front wheel and the axle center of the rear wheel respectively. The quantities can be, therefore, calculated by the following equations.

$$r_{c\_front} = R_{W\_front} - r_{T front} \times (1 - \cos\alpha) \tag{1.5}$$

$$JB = \sqrt{(JB_O)^2 - \left(R_{W\_front} - R_{W\_rear}\right)^2} \tag{1.6}$$

$$BG' = BG_0 = \sqrt{\left(Y_{G\_ini} - R_{W\_rear}\right)^2 - \left(X_{Grear\_ini}\right)^2} \tag{1.7}$$

$$\varepsilon_{01} = \varepsilon'_{01} + \varepsilon''_{01} = \arcsin\left(\frac{R_{W\_rear} - r_{c\_front}}{JB}\right) + \arctan\left(\frac{R_{W\_front} - R_{W\_rear}}{JB_O}\right) \tag{1.8}$$

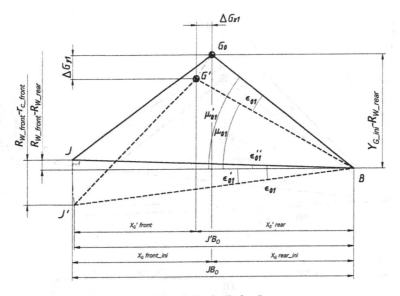

Fig. 1.7 Sway angle of the front wheel (rear wheel axis fixed)

$$\mu_{01} = \mu'_{01} - \varepsilon_{01} = \arctan\left(\frac{Y_{G\_ini} - R_{W\_rear}}{X_{Grear\_ini}}\right) - \varepsilon_{01} \qquad (1.9)$$

$$\Delta Gx_1 = BG' \times \cos\mu_{01} - X_{Grear\_ini} \qquad (1.10)$$

$$\Delta Gy_1 = \left(Y_{G\_ini} - R_{W\_rear}\right) - BG' \times \sin\mu_{01} \qquad (1.11)$$

Therefore the three main quantities referred to the sway of the front wheel can be calculated as follows.

$$J'B_O = JB \times \cos\left[\arcsin\left(\frac{R_{W\_rear} - r_{c\_front}}{JB}\right)\right]$$
$$X'_{Grear} = X_{Grear\_ini} + \Delta Gx_1 \qquad (1.12)$$
$$Y'_G = Y_{G\_ini} - \Delta Gy_1$$

Then, the sway of the rear wheel is considered keeping the axle of the front wheel fixed. The geometric relationships are indicated in Fig. 1.8 and their quantities can be calculated by the following equations.

$$r_{c\_frear} = R_{W\_rear} - r_{Trear} \times (1 - \cos\alpha) \qquad (1.13)$$

$$J'B = JB \qquad (1.14)$$

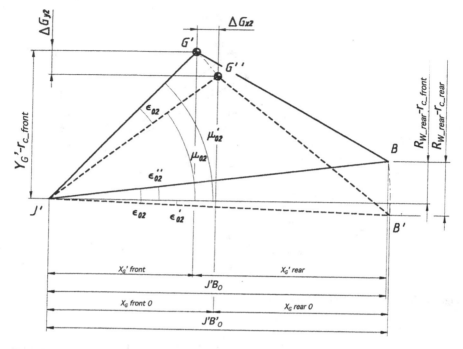

Fig. 1.8 Sway angle of the rear wheel (front wheel axis fixed)

$$J'G' = \sqrt{\left(Y'_G - r_{c\_front}\right)^2 + \left(J'B_O - X'_{Grear}\right)^2} \qquad (1.15)$$

$$\varepsilon_{02} = \varepsilon'_{02} + \varepsilon''_{02} = \arcsin\left(\frac{r_{c\_front} - r_{c\_rear}}{J'B}\right) + \arctan\left(\frac{R_{W\_rear} - r_{c\_front}}{J'B_O}\right) \qquad (1.16)$$

$$\mu_{02} = \mu'_{02} - \varepsilon_{02} = \arctan\left(\frac{Y'_G - r_{c\_front}}{J'B_O - X'_{Grear}}\right) - \varepsilon_{02} \qquad (1.17)$$

$$\Delta Gx_2 = J'G' \times \cos\mu_{02} - \left(J'B_O - X'_{Grear}\right) \qquad (1.18)$$

$$\Delta Gy_2 = Y'_G - r_{c\_front} - J'G' \times \sin\mu_{02} \qquad (1.19)$$

Therefore the three main quantities referred to the center of gravity position and due to the sway of the rear wheel can be calculated as follows.

$$J'B'_O = J'B \times \cos\varepsilon'_{02} = JB \times \cos\left[\arcsin\left(\frac{r_{c\_front} - r_{c\_rear}}{J'B}\right)\right]$$

$$X_{Grear0} = J'B'_O - J'B_O + X'_{Grear} - \Delta Gx_2 \qquad (1.20)$$

$$Y_{G0} = Y'_G - \Delta Gy_2$$

The initial configuration of the motorbike, calculated as a function of the roll angle α, is given by the final position of points G'', J' and B' of Fig. 1.8, then the center of gravity position can be modified by the diving and the rising of suspensions analyzed in the next section.

1.5 Diving and Rising of Suspensions

Equilibrium equations 1.1 and 1.2 could be directly solved if the center of gravity position was independent from the diving and rising of suspensions or if they were independent from the masses distribution.

Unfortunately the movement of suspensions depends on their geometrical, mechanical and kinematic parameters (often very complex to define especially in case of a rear suspension equipped with a progressive stiffness mechanism) and on the masses distribution, which depends on the center of gravity position. Thus the equations must be solved applying a numerical and iterative process which can be implemented into a calc sheet.

The different contributions of the front and of the rear suspensions have to be considered separately as indicated in Figs. 1.9 and 1.11.

Concerning the front suspension and referring to Fig. 1.9 it is possible to calculate the G^{III} position indicated in Fig. 1.10, as a function of the diving of the front suspension D_{front}. Since the starting configuration of Fig. 1.9 is equal to the final configuration of Fig. 1.8 the main quantities indicated in Fig. 1.9 can be calculated by the following equations.

$$J'J^* = J_1 J_1^* = r_{c\_front} - r_{c\_rear} \tag{1.21}$$

$$\lambda = \arctan\left(\frac{r_{c\_front} - r_{c\_rear}}{J'B_O'}\right) \tag{1.22}$$

$$F_1 J' = J'B' \times \sin\left(\alpha_{front} + \varepsilon_{02} - \varepsilon_{01} - \lambda\right) \tag{1.23}$$

$$F_1' J_1 = F_1 J' - D_{front} \tag{1.24}$$

$$F_1' B' = F_1 B' = J'B' \times \cos\left(\alpha_{front} + \varepsilon_{02} - \varepsilon_{01} - \lambda\right) \tag{1.25}$$

$$J_1 B' = \sqrt{\left(F_1' J_1\right)^2 + \left(F_1' B'\right)^2} \tag{1.26}$$

$$\lambda' = \arcsin\left(\frac{J_1 J_1^*}{J_1 B'}\right) = \arcsin\left(\frac{r_{c\_front} - r_{c\_rear}}{J_1 B'}\right) \tag{1.27}$$

Since $J'B' = J'B = JB$ and $\left(\alpha_{front} + \varepsilon_{02} - \varepsilon_{01}\right)' = \lambda' + \arctan\left(\frac{F_1' J_1}{F_1' B'}\right)$, it is possible to write the following formulae.

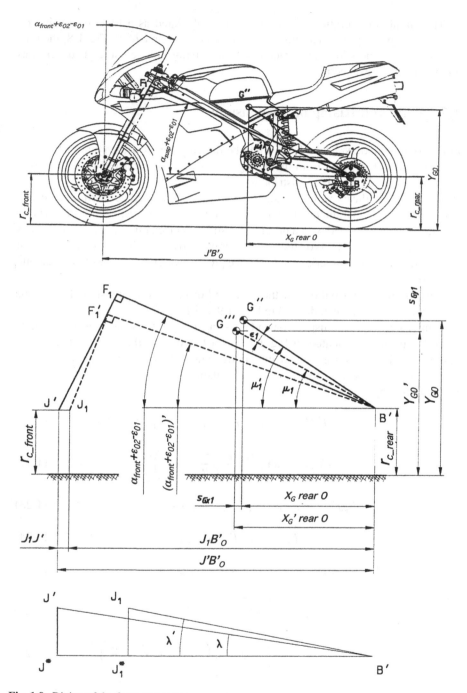

Fig. 1.9 Diving of the front suspension

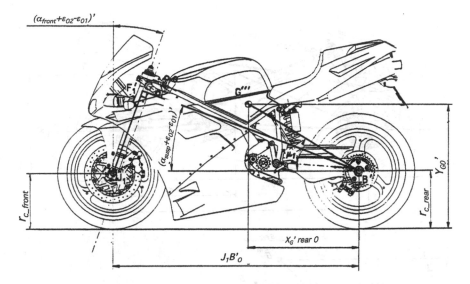

Fig. 1.10   Total sway of the front suspension

$$\varepsilon_1 = \alpha_{front} + \varepsilon_{02} - \varepsilon_{01} - \left(\alpha_{front} + \varepsilon_{02} - \varepsilon_{01}\right)' \tag{1.28}$$

$$\mu_1 = \mu_1' - \varepsilon_1 = \arctan\left(\frac{Y_{G0} - r_{c\_rear}}{X_{Grear0}}\right) - \varepsilon_1 \tag{1.29}$$

$$B'G''' = B'G'' = BG_0 = \sqrt{X_{Grear0}^2 + \left(Y_{G0} - r_{c\_rear}\right)^2} \tag{1.30}$$

$$s_{Gx1} = B'G''' \times \cos\mu_1 - X_{Grear0} \tag{1.31}$$

$$s_{Gy1} = Y_{G0} - r_{c\_rear} - B'G''' \times \sin\mu_1 \tag{1.32}$$

Therefore the three main quantities referred to the center of gravity position G^{III} and due to the diving of the front suspension, are indicated in the following system of equations.

$$\begin{aligned} J_1 B'_O &= J_1 B' \times \cos\lambda' \\ X'_{Grear0} &= X_{Grear0} + s_{Gx1} \\ Y'_{G0} &= Y_{G0} - s_{Gy1} \end{aligned} \tag{1.33}$$

The diving analysis of the rear suspension starts from the final position calculated for the front suspension and indicated in Fig. 1.10. Referring to Fig. 1.11 it is possible to define the following quantities (Fig. 1.12).

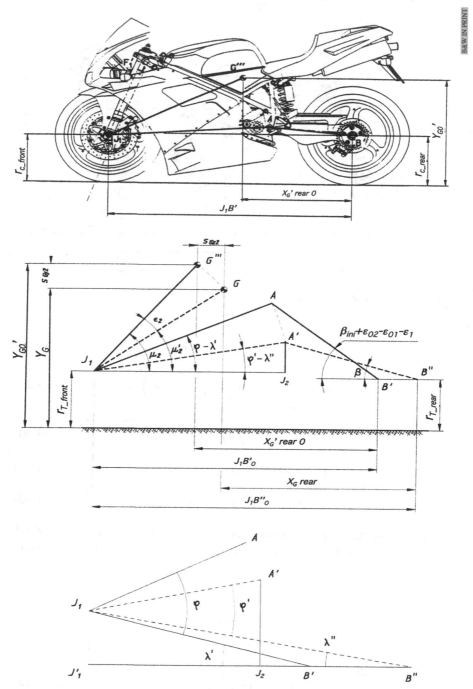

Fig. 1.11 Diving of the rear suspension

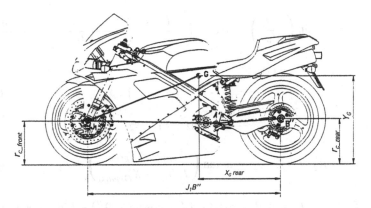

Fig. 1.12 Total sway of the rear suspension

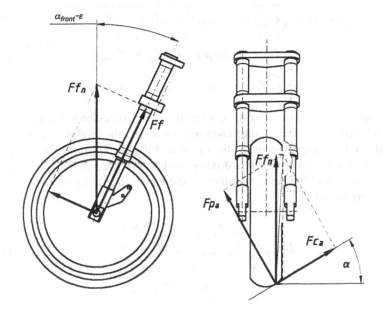

Fig. 1.13 Equilibrium of dived front suspension

$$AJ_1 = \sqrt{(AB')^2 + (J_1B')^2 - 2 \times AB' \times J_1B' \times \cos{(\beta_{ini} + \varepsilon_{02} - \varepsilon_{01} - \varepsilon_1 - \lambda')}}$$
(1.34)

$$\varphi = \arccos\left(\frac{(AJ_1)^2 + (J_1B')^2 - (AB')^2}{2 \times AJ_1 \times J_1B'}\right)$$
(1.35)

$$\left(\varphi' - \lambda''\right) = \arcsin\left(\frac{A'J_2}{A'J_1}\right) = arcsin\left(\frac{AB' \times \sin\beta - \left(r_{c\_front} - r_{c\_rear}\right)}{AJ_1}\right)$$

$$\tag{1.36}$$

$$\varepsilon_2 = \varphi - \lambda' - \left(\varphi' - \lambda''\right) \tag{1.37}$$

$$\varepsilon = \varepsilon_{01} - \varepsilon_{02} + \varepsilon_1 - \varepsilon_2 \tag{1.38}$$

$$\mu_2' = \mu_2 - \varepsilon_2 = \arctan\left(\frac{Y_{G0}' - r_{c\_front}}{J_1 B_O' - X_{Grear0}'}\right) - \varepsilon_2 \tag{1.39}$$

$$B'B'' = B''J_1' - B'J_1' = A'B'' \times \cos\beta + AJ_1 \times \cos\left(\varphi' - \lambda''\right) - J_1 B' \times \cos\lambda' \tag{1.40}$$

$$G'''J_1 = \sqrt{\left(Y_{G0}' - r_{c\_front}\right)^2 + \left(J_1 B_O' - X_{Grear0}'\right)^2} \tag{1.41}$$

$$s_{Gx2} = G'''J_1 \times \cos\mu_2' - \left(J_1 B_O' - X_{Grear0}'\right) \tag{1.42}$$

$$s_{Gy2} = Y_{G0}' - r_{c\_front} - G'''J_1 \times \sin\mu_2' \tag{1.43}$$

The angles ε are considered positive when the front suspension is diving more than the rear one (the rotation is counterclockwise) whereas the angles β_{ini} of Eq. 1.34 and β of Eq. 1.36 are, respectively, the initial and the current positions of the rear suspension depending on the actual driving conditions (speed, roll angle and center of gravity position). The current position of the center of gravity is, then, given by the following equation.

$$J_1 B_O'' = J_1 B_O' + B'B''$$
$$X_{Grear} = X_{Grear0}' - s_{Gx2} + B'B'' \tag{1.44}$$
$$X_{Gfront} = J_1 B_O'' - X_{Grear}$$
$$Y_G = Y_{G0}' - s_{Gy2}$$

The diving value of the front suspension D_{front} used in Eq. 1.24, can be calculated by means of Eq. 1.45 as a function of F_{pa}, F_{ca}, the stiffness of the front suspension K_{front}, the preloading step D_{ifront} and the current caster angle indicated in Fig. 1.13. This function is the first implicit dependance of equilibrium equations, even if the current diving of the front suspension can be automatically adjusted, for example, by a calc sheet, when the initial position and geometrical quantities are known.

$$D_{front} = \frac{\left(F_{pa} \times \cos\alpha + F_{ca} \times \sin\alpha\right) \times \cos\left(\alpha_{front} - \varepsilon\right)}{K_{front}} - D_{ifront} \tag{1.45}$$

Unfortunately the position of the center of gravity also depends on the diving and on the rising of the rear suspension that can be expressed as a function of the β angle reported in the scheme of Fig. 1.11 and in Eq. 1.36. The β angle depends on the following parameters:

1. The mass distribution and, consequently, the position of the center of gravity.
2. The equilibrium of the rear suspension.

The first parameter represents the second implicit function to be expressed and solved by means of a calc sheet.

The second parameter has to be analysed once the architecture of the mechanism, which produces the diving and the rising of the suspension, has been extensively defined. In this book a suspension made of a single arm equipped with an articulated mechanism with a progressive stiffness will be presented and analyzed in the Chap. 2 so that an iteration process or a *goal search* tool has to be used in order to solve the problem.

References

1. Cossalter V, Lot R, Maggio F (2004) On the stability of motorcycle during braking. In: Small engine technology conference & exhibition, Graz, Austria, Sept 2004
2. Cossalter V (2006) Motorcycle dynamics. LULU Publisher, Modena

Chapter 2
The Rear Suspension Equilibrium

Abstract This chapter aims at defining the equations that govern the rear suspension in term of the overall equilibrium and loads applied for different conditions of the motorbike motion. The formulae are useful for designing a rear suspension made by a unique arm which connects the frame pin with the wheel axle. The equations are non-linear and implicit so that they can be solved by recursive methods applicable in a calc sheet. The mathematical models can be applied to different types of rear suspension but they are developed for the peculiar case of those equipped with an articulated mechanism which leads to a progressive stiffness of a single arm rear suspension.

2.1 Power and Load Transmission

The power transmitted to the rear wheel of a motorbike ridden into a straight plane or in a bend, depends on the *power curve* of the engine reported, for example, in Fig. 2.1; the engine thrust is, obviously, able to overcome all the forces which are opposed to the motorbike motion. If the gear ratio and the motorbike speed are known, it is possible to calculate the torque moment applied to the rear wheel by means of the following equation in which W is the power and ω is the current speed of the rear wheel.

$$T = \frac{W}{\omega} = \frac{W \times r_{crear}}{v} \tag{2.1}$$

The final transmission of the power is, usually, produced by a chain that acts with a force F_{chain} on the chain crown reported in Fig. 2.2. The traction moment transmitted to the wheel produces the traction force F_t through the grip of the tyre with the ground; these forces can be calculated with the following equations.

$$F_{chain} = \frac{2 \times T}{\phi_c} = \frac{2 \times W \times r_{crear}}{\phi_{c \times v}} \tag{2.2}$$

D. Croccolo and M. De Agostinis, *Motorbike Suspensions*,
SpringerBriefs in Applied Sciences and Technology,
DOI: 10.1007/978-1-4471-5149-4_2, © The Author(s) 2013

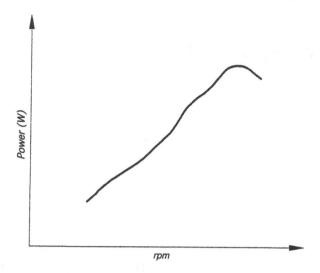

Fig. 2.1 Power versus revolution curve for the endothermic engine

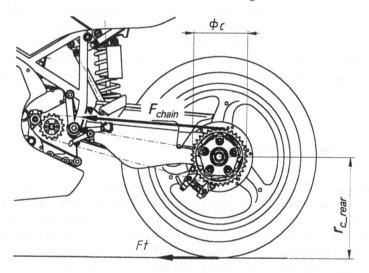

Fig. 2.2 Scheme of forces and reactions acting on the rear wheel

$$F_t = \frac{W}{v} \qquad (2.3)$$

The chain load F_{chain} and the traction load F_t as mentioned above, depend on the maximum power transmitted and on the motorbike speed; anyway both of them depend on the grip limit of the rear wheel and on the contact limit of the front wheel (the capsizing limit of the motorbike). The best way to consider the actual load

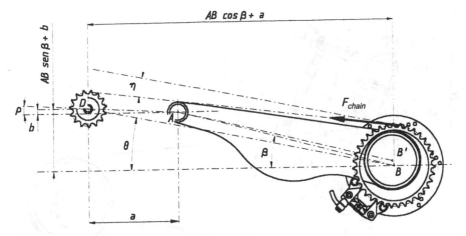

Fig. 2.3 Rear suspension scheme

depending on the motorbike speed, is to choose the lowest value of the three loads generated by the maximum power, the grip limit and the capsizing limit.

Once the values of F_{chain} and F_t are defined it is necessary to apply them to the rear suspension. While the direction of F_t is known the action line of F_{chain} is much more complicated to define. Referring to the scheme of Figs. 2.3 and 2.4 F_{chain} acts on the $\theta - \eta$ direction since the point B' is moved from its initial position by the rotation θ_7 of an eccentric tool with a gap $e = BB'$ necessary for the initial tightening of the chain (see the quantities of Fig. 2.4).

Once the initial abscissa a_{ini} and ordinate b_{ini} of point D respect to point A and the diameters of pinion ϕ_P and of crown ϕ_C reported in Fig. 2.3 are known, it is possible to obtain the following equations.

$$AB = \sqrt{(AB' - e \times \cos\theta_7)^2 + (e \times \sin\theta_7)^2} \qquad (2.4)$$

$$a = \sqrt{a_{ini}^2 + b_{ini}^2} \times \cos\left[\arctan\left(\frac{b_{ini}}{a_{ini}}\right) - \varepsilon\right] \qquad (2.5)$$

$$b = \sqrt{a_{ini}^2 + b_{ini}^2} \times \sin\left[\arctan\left(\frac{b_{ini}}{a_{ini}}\right) - \varepsilon\right] \qquad (2.6)$$

$$\theta = \arctan\left(\frac{AB \times \sin\beta + b}{AB \times \cos\beta + a}\right) \qquad (2.7)$$

$$\eta = \arcsin\left[\frac{(\phi_C - \phi_P) \times \cos\theta}{2 \times (AB \times \cos\beta + a)}\right] \qquad (2.8)$$

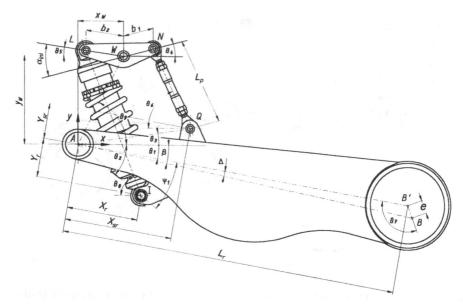

Fig. 2.4 Rear arm scheme and representation

2.2 Forces Acting on the Suspension Mechanism

The rear suspension under investigation is equipped with an articulated mechanism that produces the progression of its stiffness. The articulated mechanism represented in Fig. 2.4 and zoomed in Fig. 2.5, is analysed starting from the following vectorial equation.

$$\overline{AQ} + \overline{QN} + \overline{NW} + \overline{WA} = 0 \tag{2.9}$$

The vectorial equation 2.9 is rewritten into the two scalar formulae of Eq. 2.10.

$$AQ \times \sin\theta_3 + QN \times \sin\theta_6 - NW \times \sin\theta_4 - Y_W = 0$$
$$AQ \times \cos\theta_3 - QN \times \cos\theta_6 - NW \times \cos\theta_4 - X_W = 0 \tag{2.10}$$

By substituting θ_6 in formulae 2.10, a function between θ_3 and θ_4 can be obtained and reported in Eq. 2.11, whereas A, B, C, θ_3, Δ, X_W and Y_W are expressed in the formulae of Eq. 2.12 (X_{Wini} and Y_{Wini} are the initial coordinates). The β angle of Eq. 2.12 represents the current position of the rear suspension depending on the riding conditions.

$$A \times \sin\theta_4 + B \times \cos\theta_4 = C \tag{2.11}$$

$$X_W = \sqrt{X_{Wini}^2 + Y_{Wini}^2} \times \cos\left[\arctan\left(\frac{Y_{Wini}}{X_{Wini}}\right) + \varepsilon\right]$$

$$Y_W = \sqrt{X_{Wini}^2 + Y_{Wini}^2} \times \sin\left[\arctan\left(\frac{Y_{Wini}}{X_{Wini}}\right) + \varepsilon\right]$$

$$\Delta = \arctan\left(\frac{B'B \times \sin\theta_7}{AB' - B'B \times \cos\theta_7}\right)$$

$$\theta_1 = \beta - \Delta$$

$$\theta_2 = \theta_1 + \arctan\left(\frac{Y_r}{X_r}\right) \qquad (2.12)$$

$$\theta_3 = \arctan\left(\frac{Y_{sr}}{X_{sr}}\right) - (\beta - \Delta)$$

$$A = 2 \times Y_W \times b_1 - 2 \times b_1 \times \sqrt{X_{sr}^2 + Y_{sr}^2} \times \sin\theta_3$$

$$B = 2 \times X_W \times b_1 - 2 \times b_1 \times \sqrt{X_{sr}^2 + Y_{sr}^2} \times \cos\theta_3$$

$$C = 2 \times \sqrt{X_{sr}^2 + Y_{sr}^2} \times (Y_W \times \sin\theta_3 + X_W \times \cos\theta_3) +$$
$$- b_1^2 - X_{sr}^2 - Y_{sr}^2 - X_W^2 - Y_W^2 + NQ^2$$

The Eq. 2.11 can be solved since $sin\theta_4 = \frac{2 \times \tan\left(\frac{\theta_4}{2}\right)}{1 + \tan^2\left(\frac{\theta_4}{2}\right)}$ and $cos\theta_4 = \frac{1 - \tan^2\left(\frac{\theta_4}{2}\right)}{1 + \tan^2\left(\frac{\theta_4}{2}\right)}$. The acceptable solution for θ_4 is given by Eq. 2.13, therefore by means of the formulae of Eq. 2.14 it is possible to calculate the diving of the spring rod (D_{rear}) as indicated by Eq. 2.15.

$$\theta_4 = 2 \times \arctan\left(\frac{A - \sqrt{A^2 + B^2 - C^2}}{B + C}\right) \qquad (2.13)$$

$$\theta_5 = \alpha_{ini} - \theta_4$$

$$\theta_6 = \arcsin\left(\frac{Y_W + b_1 \times \sin\theta_4 - \sqrt{X_{sr}^2 + y_{sr}^2} \times \sin\theta_3}{L_p}\right) \qquad (2.14)$$

$$\theta_8 = \arctan\left(\frac{Y_W + b_2 \times \sin\beta_5 + \sqrt{X_r^2 + Y_r^2} \times \sin\theta_2}{\sqrt{X_r^2 + Y_r^2} \times \cos\theta_2 - X_W + b_2 \times \cos\theta_5}\right)$$

$$D_{rear} = \Delta D_{irear} - \sqrt{\frac{\left(\sqrt{X_r^2 + Y_r^2} \times \cos\theta_2 - X_W + b_2 \times \cos\theta_5\right)^2}{+ \left(Y_W + b_2 \times \sin\theta_5 + \sqrt{X_r^2 + Y_r^2} \times \sin\theta_2\right)^2}} \qquad (2.15)$$

ΔD_{irear} is the initial elongation of the spring rod when the rear suspension is completely dived up (the β angle is equal to a β_{ini} value).

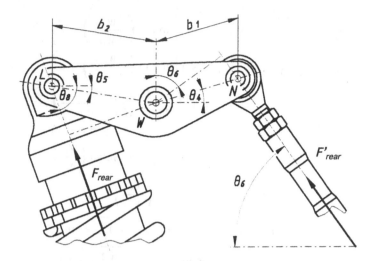

Fig. 2.5 Zoom on the superior part of the articulated mechanism of the rear suspension

Referring to Fig. 2.5 it is, finally, possible to calculate the forces generated by the rear suspension F_{rear} and F'_{rear} once the preloading step (D_{irear}) and the stiffness (K_{rear}) of the rear suspension are known.

$$F_{rear} = (D_{rear} + D_{irear}) \times K_{rear} \tag{2.16}$$

$$F'_{rear} = F_{rear} \times \frac{b_2 \times \sin(\theta_8 - \theta_5)}{b_1 \times \sin(\theta_4 + \theta_6)} \tag{2.17}$$

2.3 Braking and Traction Forces

The maximum braking force and the maximum traction force must be considered acting separately on the motorbike so that during the braking phase the thrust of the engine is equal to zero (no force of traction is applied to the power chain and to the rear wheel), whereas during the maximum thrust of the engine the braking forces have to considered equal to zero.

The maximum braking force acting on the rear suspension must be produced when the motorbike is ridden into a straight plane because the bend reduces the grip of tyres and, therefore, the braking force. Furthermore the front wheel is supposed to be free so that the entire braking power is considered to be applied to the rear tyre.

The study starts from the equilibrium formulae presented in Chap. 1 and here reported.

$$Fp_a = \frac{Fp \times (X_Grear + f_p \times Y_G \times \cos\alpha) - Y_G \times \cos\alpha \times (Ft - Fb)}{X_Gfront + X_Grear - Y_G \times \cos\alpha \times (f_a - f_p)}$$

$$Fp_p = \frac{Fp \times (X_Gfront - f_a \times Y_G \times \cos\alpha) + Y_G \times \cos\alpha \times (Ft - Fb)}{X_Gfront + X_Grear - Y_G \times \cos\alpha \times (f_a - f_p)} \quad (2.18)$$

$$Fc_a = Fp_a \times \tan\alpha$$

$$Fc_p = Fp_p \times \tan\alpha$$

$$\alpha = \arctan\frac{v^2}{r \times g}$$

Since the grip limit of the rear wheel has to be considered during the braking phase, it is necessary to add the following condition.

$$\sqrt{(F_b + Ar_p - F_t)^2 + (Fc_p)^2} \le \mu_g \times Fp_p \quad (2.19)$$

Therefore the maximum braking force with $Ar_p = Fp_p \times f_p$ is given by Eq. 2.20.

$$F_{bmax} = F_t - Fp_p \times f_p + \sqrt{(\mu_g \times Fp_p)^2 - (Fc_p)^2} \quad (2.20)$$

Since the curve radius is equal to ∞ and the $Fc = 0$ Eq. 2.18 becomes Eq. 2.21 while Eq. 2.20 becomes Eq. 2.22.

$$Fp_a = \frac{Fp \times (X_Grear + f_p \times Y_G \times \cos\gamma) - Y_G \times \cos\gamma \times (Ft - Fb)}{X_Gfront + X_Grear - Y_G \times \cos\gamma \times (f_a - f_p)}$$

$$Fp_p = \frac{Fp \times (X_Gfront - f_a \times Y_G \times \cos\gamma) + Y_G \times \cos\gamma \times (Ft - Fb)}{X_Gfront + X_Grear - Y_G \times \cos\gamma \times (f_a - f_p)} \quad (2.21)$$

$$Fc_a = 0$$

$$Fc_p = 0$$

$$\alpha = 0$$

$$F_{bmax} = F_t + Fp_p \times (\mu_g - f_p) \quad (2.22)$$

Now by substituting Eq. 2.21 into Eq. 2.22 it is possible to obtain Eq. 2.23 in which F_t is equal to zero.

$$F_{bmax} = Fp \times \frac{(\mu_g - f_p) \times (X_Gfront - f_a \times Y_G \times \cos\gamma)}{X_Gfront + X_Grear + Y_G \times \cos\gamma \times (\mu_g - f_a)} \quad (2.23)$$

The F_t load produces its effect on the rear suspension as indicated in the scheme of Fig. 2.6. The braking disk applies the F_d reaction to the disk frame which is fixed to the rear suspension by means of a pin where the braking force is constrained to the suspension by means of F_{pin} reaction on the disk frame. F_d and F_{pin} are calculated by Eqs. 2.24 and 2.25.

$$F_d = F_b \times \frac{r_{crear}}{r_d} \tag{2.24}$$

$$F_{pin} = F_d \times \frac{r_d}{BC \times \cos\theta_{12}} \tag{2.25}$$

F_{pin} acts on the normal to the direction CB' of Fig. 2.6; in Eq. 2.26 are reported the missing parameters necessary for calculation.

$$BC = \sqrt{L_{pin}^2 + (B'B)^2 - 2 \times L_{pin} \times BB' \times \cos\theta_7}$$

$$\theta_{10} = \arccos\left[\frac{L_{pin}^2 + (BC)^2 - (B'B)^2}{2 \times L_{pin} \times BC}\right]$$

$$\theta_{11} = 90 - (\chi + \theta_{12} - \theta_{10}) \tag{2.26}$$

$$\theta_{12} = \arcsin\left(\frac{h_{pin}}{BC}\right)$$

$$\theta_{13} = \theta_{12} - \theta_{10}$$

The braking forces acting on the wheel axle of the rear suspension are given by the F_{pin} force and by the F_d force applied to the frame of the braking clamps and by the F_b indicated in Fig. 1.2 of Chap. 1. Concerning the forces applied to the frame (F_d and F_{pin}) they are supported by the axle bearings indicated in Fig. 2.8. Since the stresses acting on the rear suspension depends on the bearing reactions (R_{b11}, R_{b12}, R_{b21} and R_{b22}), it is necessary to impose the overall equilibrium by means of the formulae of Eq. 2.27 in order to calculate the stresses defined by Eqs. 2.28, 2.29, 2.30 and 2.31. The symbols used in Eq. 2.27 are referred to Figs. 2.7, 2.8 and 2.10.

$$- R_{b21} + F_{pin} \times \sin\theta_{12} + R_{b11} - F_d \cos(\theta_{11} - \theta_{10}) = 0$$

$$-R_{b22} + F_{pin} \times \cos\theta_{12} - R_{b12} - F_d \sin(\theta_{11} - \theta_{10}) = 0$$

$$-R_{b22} \times (d_2 - d_1) + F_{pin} \times \cos\theta_{12} \times d_4 + F_d \sin(\theta_{11} - \theta_{10}) \times d_5 = 0 \tag{2.27}$$

$$R_{b21} \times (d_2 - d_1) - F_{pin} \times \sin\theta_{12} \times d_4 - F_d \cos(\theta_{11} - \theta_{10}) \times d_5 = 0$$

The solutions of the system of Eq. 2.27 are indicated in the following equations.

$$R_{b11} = \frac{F_d \cos(\theta_{11} - \theta_{10}) \times (d_2 - d_1 + d_5) - F_{pin} \times \sin\theta_{12} \times (d_2 - d_1 - d_4)}{d_2 - d_1} \tag{2.28}$$

$$R_{b12} = \frac{F_{pin} \times \cos\theta_{12} \times (d_2 - d_1 - d_4) - F_d \sin(\theta_{11} - \theta_{10}) \times (d_2 - d_1 + d_5)}{d_2 - d_1} \tag{2.29}$$

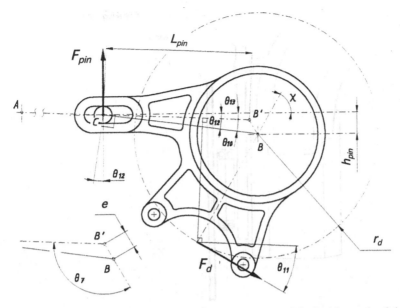

Fig. 2.6 Scheme of the external reactions applied to the frame of the braking pads of the rear suspension

$$R_{b21} = \frac{F_d \cos(\theta_{11} - \theta_{10}) \times d_5 + F_{pin} \times \sin\theta_{12} \times d_4}{d_2 - d_1} \qquad (2.30)$$

$$R_{b22} = \frac{F_d \sin(\theta_{11} - \theta_{10}) \times d_5 + F_{pin} \times \cos\theta_{12} \times d_4}{d_2 - d_1} \qquad (2.31)$$

Since it is useful to align the loads to the normal n and to the parallel p directions with the respect to the rear suspension axis (AB' of Fig. 2.4), referring the scheme of Fig. 2.9, it is possible to write the following equations.

$$R_{bp1} = R_{b11} \times \cos\theta_{10} + R_{b12} \times \sin\theta_{10} \qquad (2.32)$$

$$R_{bn1} = R_{b12} \times \cos\theta_{10} - R_{b11} \times \sin\theta_{10} \qquad (2.33)$$

$$R_{bp2} = R_{b21} \times \cos\theta_{10} - R_{b22} \times \sin\theta_{10} \qquad (2.34)$$

$$R_{bn2} = R_{b22} \times \cos\theta_{10} + R_{b21} \times \sin\theta_{10} \qquad (2.35)$$

The braking force F_b reported in Fig. 1.2 of Chap. 1 and the traction forces F_{chain} and F_t indicated in Fig. 2.2 have to be analyzed with the same procedure presented above which leads to the definition of reactions acting on the bearings of the rear suspension.

The following formulation is made referring to the scheme of Fig. 2.10. As mentioned before the traction forces and the braking force do not act at the same time: when considering the former the latter is imposed equal to zero and vice versa.

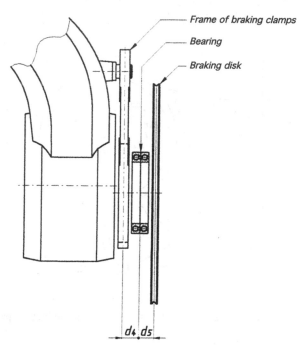

Fig. 2.7 Scheme of the braking disk and of the pads frame of the rear suspension

Furthermore, as indicated in the scheme of Fig. 2.10, only the right bearing is able to support the axial force R_{13}.

The overall equilibrium of the forces reported in Fig. 2.10 can be written in the formulae of Eq. 2.36.

$$F_{chain} \times \cos(\theta - \eta) - R_{21} + R_{11} + Ft - F_b - f_p \times Fp_p = 0$$
$$F_{chain} \times \sin(\theta - \eta) - R_{22} - R_{12} + Fc_p \times \sin\alpha + Fp_p \times \cos\alpha = 0$$
$$R_{13} - Fc_p \times \cos\alpha + Fp_p \times \sin\alpha = 0$$
$$F_{chain} \times \sin(\theta - \eta) \times d_3 - R_{22} \times d_2 - R_{12} \times d_1 + Fc_p \times \cos\alpha \times r_{crear} +$$
$$-Fp_p \times \sin\alpha \times r_{crear} + (Fc_p \times \sin\alpha + Fp_p \times \cos\alpha) \times CK_{rear} = 0 \quad (2.36)$$
$$-F_{chain} \times \cos(\theta - \eta) \times d_3 + R_{21} \times d_2 +$$
$$-R_{11} \times d_1 + (F_b - Ft + f_p \times Fp_p) \times CK_{rear} = 0$$

The solutions of the system of Eq. 2.36 are indicated in the following equations.

$$R_{11} = \frac{F_{chain} \times \cos(\theta - \eta) \times (d_3 - d_2) - (Ft - F_b - f_p \times Fp_p) \times (d_2 - CK_{rear})}{d_2 - d_1}$$

$$R_{12} = \frac{Fc_p \times [(d_2 - CK_{rear})\sin\alpha - r_{crear}\cos\alpha] - F_{chain} \times \sin(\theta - \eta) \times (d_3 - d_2)}{d_2 - d_1} +$$

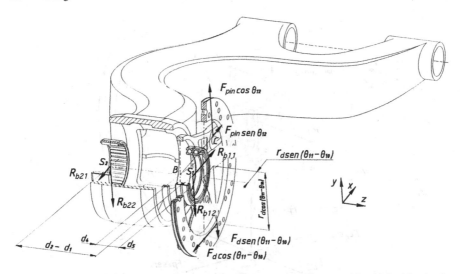

Fig. 2.8 Scheme of the bearings reactions of the rear suspension generated by the braking loads

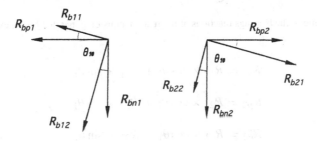

Fig. 2.9 Components of the bearings reactions generated by the braking loads

$$+\frac{Fp_p \times [(d_2 - CK_{rear}) \cos\alpha + rc_{rear} \sin\alpha]}{d_2 - d_1}$$

$$R_{13} = Fc_p \times \cos\alpha - Fp_p \times \sin\alpha \quad (2.37)$$

$$R_{21} = \frac{F_{chain} \times \cos(\theta - \eta) \times (d_3 - d_1) - (Ft - F_b - f_p \times Fp_p) \times (d_1 - CK_{rear})}{d_2 - d_1}$$

$$R_{22} = \frac{Fc_p \times [r_{crear} \cos\alpha - (d_1 - CK_{rear}) \sin\alpha] + F_{chain} \times \sin(\theta - \eta) \times (d_3 - d_1)}{d_2 - d_1} +$$

$$-\frac{Fp_p \times [r_{crear} \sin\alpha + (d_1 - CK_{rear}) \cos\alpha]}{d_2 - d_1}$$

As described above is useful to align the loads to the normal n and to the parallel p directions with the respect to the rear suspension axis (AB' of Fig. 2.4). Referring the scheme of Fig. 2.11 it is possible to write the following equations.

$$R_{p1} = R_{11} \times \cos\theta_1 - R_{12} \times \sin\theta_1 \quad (2.38)$$

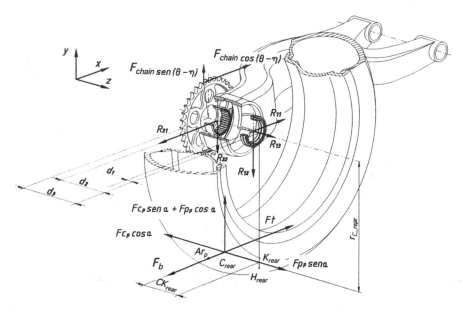

Fig. 2.10 Scheme of the bearings reactions of the rear suspension generated by the external loads

$$R_{n1} = R_{12} \times \cos\theta_1 + R_{11} \times \sin\theta_1 \qquad (2.39)$$

$$R_{p2} = R_{21} \times \cos\theta_1 + R_{22} \times \sin\theta_1 \qquad (2.40)$$

$$R_{n2} = R_{22} \times \cos\theta_1 - R_{21} \times \sin\theta_1 \qquad (2.41)$$

The total forces applied to the axle of the rear suspension are calculated by the formulae of Eq. 2.42 and reported in the scheme of Fig. 2.12.

$$\begin{aligned}
R_{t11} &= R_{bp1} + R_{p1} \\
R_{t12} &= R_{bn1} + R_{n1} \\
R_{t13} &= R_{13} \\
R_{t21} &= R_{bp2} + R_{p2} \\
R_{t22} &= R_{bn2} + R_{n2}
\end{aligned} \qquad (2.42)$$

2.4 Global Forces Acting on Rear Suspension

Referring to the scheme of Fig. 2.13, in which is applied the global system of forces acting on the rear suspension, it is, now, possible to define the rotating equilibrium of the rear suspension by means of Eq. 2.43.

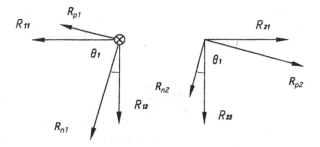

Fig. 2.11 Components of the bearings reactions generated by the external loads

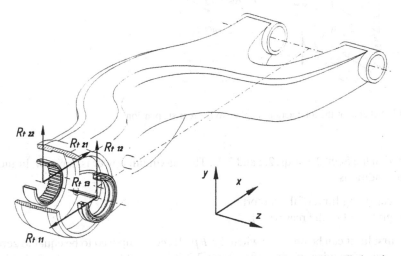

Fig. 2.12 Scheme of the bearings actions generated by the total loads and applied to the axle of the rear suspension

$$
\begin{aligned}
&F'_{rear} \times \sin(\theta_6 - \theta_1) \times X_{sr} + F'_{rear} \times \cos(\theta_6 - \theta_1) \times Y_{sr} \\
&- F_{rear} \times \cos(\theta_8 - \theta_1) \times Y_r + F_{rear} \times \sin(\theta_8 - \theta_1) \times X_r \\
&- R_{t12} \times (L_r - e \times \cos\theta_7) - R_{t11} \times e \times \sin\theta_7 - R_{t22} \times (L_r - e \times \cos\theta_7) \\
&+ R_{t21} \times e \times \sin\theta_7 + F_{pin} \times \cos\theta_{13} \times (L_r - L_{pin}) = 0 \qquad (2.43)
\end{aligned}
$$

Equation 2.43 is a recursive equation since it has to be solved starting by an approximated configuration of the motorbike and of the rear suspension and adjusting the parameters until the difference between the left and the right terms is lower than a fixed value: essentially the tendency of the left term has to be imposed equal to the null value in the calc sheet.

The loads applied to the rear suspension depend on the traction force F_t, on the chain force F_{chain} and on the braking force F_b. Traction forces (F_t and F_{chain}) depend on the maximum power provided by the engine at a stated speed, as well

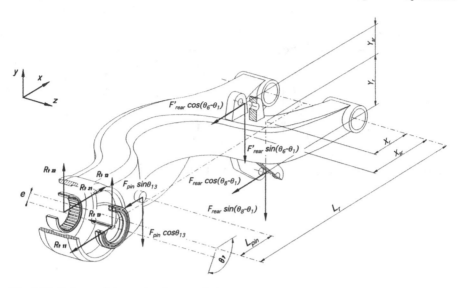

Fig. 2.13 Scheme of the total actions applied to the rear suspension

indicated in the Sect. 2.1 (Eqs. 2.2 and 2.3). The maximum power is, anyway, limited by two conditions:

1. The capsizing limit of the motorbike.
2. The grip limit of the rear wheel.

The first limit can be obtained when the Fp_a force is imposed to be equal to zero. Recalling the formulae of Eq. 2.18 of Sect. 2.3 it is possible to write the following equation:

$$Fp \times (X_G rear + f_p \times Y_G \times \cos\alpha) - Y_G \times \cos\alpha \times (Ft_{max} - Fb) = 0 \quad (2.44)$$

Therefore the maximum power transmitted at the capsizing limit W_{maxc} is given by:

$$W_{maxc} = v \times \left[\frac{Fp \times (X_G rear + f_p \times Y_G \times \cos\alpha)}{Y_G \times \cos\alpha} + Fb \right] \quad (2.45)$$

The grip limit for the rear wheel can be expressed by the Eq. 2.46 once the coefficient of friction μ is known.

$$Fp_p \times \mu \geq \sqrt{\left(Ft - Fb - Ar_p\right)^2 + \left(Fc_p\right)^2} \quad (2.46)$$

Since $Ar_p = Fp_{pa}$, the maximum power transmitted at the grip limit W_{maxg} is given by:

$$W_{maxg} = v \times \left[Fp \times f_p + \sqrt{(Fp_p \times \mu)^2 - (Fc_p)^2} + Fb \right] \qquad (2.47)$$

Finally the front wheel grip limit has to be controlled and verified by Eq. 2.48. This limit is independent by the power but is referred to the centripetal acceleration so that this limit can be considered for the highest roll angles.

$$Fp_a \times \sqrt{\mu^2 - f_a^2} \geq Fc_a \qquad (2.48)$$

In conclusion, loads and maximum power imposed for a stated speed must respect both the capsizing and the grip limit of the rear wheel whereas the grip limit of the front wheel must be controlled when the roll angle is different from the zero value.

The global loads system acting on the rear suspension depends, essentially, on the speed value, on the roll angle, on the braking force and on the traction force. In order to take into account the different combinations of the aforementioned factors the best way to proceed is the use a calc sheet. The whole roll angle positions have to be analyzed considering their different grip limit and traction limit for a range of imposed speeds; by doing this it is possible to simulate the different bending condition combined with the different speeds. For each combination, the outer limits of the maximum power applied will be considered limited by the grip or capsizing limits and in the absence of traction; the braking action is not considered during the bending.

The braking loads are applied in case of riding into a straight plane conducted up to the grip limit; in this case the traction loads are, obviously, kept equal to the zero value.

The aforementioned conditions are applied considering both the presence of the sole rider and in the case of transportation of one passenger. The motorbike is considered equipped with all the liquids (oil, water and petrol) fully filled into their cans.

Before presenting and discussing in Chap. 3 an example of calculation, it is necessary to highlight that during the riding into a straight plane, when the capsizing limit condition is achieved, the D_{front} parameter calculated by Eq. 1.45 of Chap. 1 must be set at the null value; this means that negative values are not admitted for the D_{front} parameter.

Chapter 3
Example of Loads and Stresses Acting on the Rear Suspension

Abstract In this chapter is presented an example (*case study*) of calculations and definitions of loads and stresses obtained by applying the formulae discussed in the Chaps. 1 and 2. The results are obtained using a calc sheet in order to solve the recursive problem caused by the implicit function of the equilibrium equation associated to the rear suspension and in order to produce the diagrams of loads and stresses. Loads and stresses have been calculated depending on some fixed and variable parameters while results are presented in tables and diagrams so that it is possible to highlight their trend and their maximum values.

3.1 Introduction

The dynamic equilibrium of a motorbike riding into a straight plane or in a bend, has been extensively analysed and discussed in the Chaps. 1 and 2. The present chapter is dedicated to the calculation of loads and stresses acting on the rear suspension; the rear suspension is formed by a unique arm and by a special articulated mechanism able to produce a stiffness progression of the suspension which is deeply presented in the Chap. 2. The loads and stresses have been calculated by means of a calc sheet in which the formulae presented in the Chaps. 1 and 2 have been applied and utilized. The use of the calc sheet is, really, effective for solving the recursive problem occurred in the equilibrium equation of the rear suspension and for producing the diagrams of loads and stresses. In order to solve the unique implicit equation the *goal search* tool provided by the calc sheet has been extensively used for determining the actual rotation angle (β_{act}) of the rear suspension.

3.2 Initial Data and Calculations

Firstly the basic quantities and data (both mechanical and geometrical) described in the Chaps. 1 and 2, have been defined and collected in Table 3.1. The initial data inserted in Table 3.1 are distinguished for the rider plus passenger loading condition

Table 3.1 Initial quantities and data of motorbike: rider and passenger without parenthesis, only rider between parenthesis

Description	Symbol	Value	Unit
Traction maximum power	W_{max}	93.2	(kW)
Total mass (driver and passenger)	m_{tot}	420 (325)	(kg)
Wheel axis distance	JB_{ini}	1,440	(mm)
Center of gravity position	X_{Grear_ini}	497 (633)	(mm)
Center of gravity position	Y_{G_ini}	727 (621)	(mm)
Front wheel radius	R_{W_front}	300	(mm)
Rear wheel radius	R_{W_rear}	310	(mm)
Front tyre radius	r_{Tfront}	57	(mm)
Rear tyre radius	r_{Trear}	115	(mm)
Roll coefficient of friction of the front wheel	f_a	0.012	
Roll coefficient of friction of the rear wheel	f_r	0.012	
Grip coefficient of friction	μ	1.4	
Caster angle	α_{front}	26.5	(°)
Preloading step of the front suspension	D_{ifront}	7	(mm)
Maximum motion of the front suspension	D_{mfront}	130	(mm)
Stiffness of the front suspension	K_{front}	12.6	(N/mm)
Preloading step of the rear suspension	D_{ifront}	25 (10)	(mm)
Maximum motion of the rear suspension	D_{mrear}	135	(mm)
Stiffness of the rear suspension	K_{front}	105	(N/mm)
Chain pitch	p_c	15.875	(mm)
Tooth number of pinion	z_p	15	
Tooth number of gear	z_c	39	
Distance between the pinion axis and rear fulcrum	a_{ini}	94.5	(mm)
Distance between the pinion axis and rear fulcrum	b_{ini}	14	(mm)
Distance between the rear fulcrum and the gear axis	L_r	565	(mm)
Rear suspension orientation	θ_{1ini}	9	(°)
Distance between the rear fulcrum and the spring	X_r	182	(mm)
Distance between the rear fulcrum and the spring	Y_r	87.5	(mm)
Distance between the rear fulcrum and the rod arm	X_{sr}	216.5	(mm)
Distance between the rear fulcrum and the rod arm	Y_{sr}	145	(mm)
Distance between the rear and the rocker arm fulcrums	X_{Wini}	39	(mm)
Distance between the rear and the rocker arm fulcrums	Y_{Wini}	163	(mm)
Rocker arm dimension	b_1	76.5	(mm)
Rocker arm dimension	b_2	60	(mm)
Rocker arm angle	α_{ini}	29	(°)
Rod dimension	L_p	166	(mm)
B'B distance	e	19	(mm)
B'B angle	θ_7	165	(°)
Distance of brake pin from the rear wheel axis	L_{pin}	120	(mm)
Distance of brake pin from the rear wheel axis	h_{pin}	18.7	(mm)
Braking disk radius	r_d	95	(mm)

(continued)

Table 3.1 (Continued)

Description	Symbol	Value	Unit
Braking disk radius	r_d	95	(mm)
Pad brakes angle	χ	75	(°)
Bearings distance on the rear wheel axle	d_1	14	(mm)
Bearings distance on the rear wheel axle	d_2	85	(mm)
Bearings distance on the rear wheel axle	d_3	123	(mm)
Bearings distance on the rear wheel axle	d_4	1	(mm)
Bearings distance on the rear wheel axle	d_5	12.5	(mm)

W_{max} [W]	93,200	L_r [mm]	565	Fp [N]	4,120
m_{tot} [kg]	420	θ_{1ini} [°]	9	$atan(\mu)$ [°]	54
$JB_{ini}=JB_0$ [mm]	1,440	Xr [mm]	182.00	Φ_p [mm]	75.798
X_{Grear_ini} [mm]	497	Yr [mm]	87.50	Φ_c [mm]	197.074
Y_{G_ini} [mm]	727	Xsr [mm]	216.50	Δ [°]	0.483
R_{W_front} [mm]	300	Ysr [mm]	145.00	β_{ini} [°]	9.483
r_{T_front} [mm]	57	α_{ini} [°]	29.00	θ_{2ini} [°]	35
R_{W_rear} [mm]	310	Xwini [mm]	39.00	θ_{3ini} [°]	25
r_{T_rear} [mm]	115	Ywini [mm]	163.00	A_{ini} [mm]	8,209
fa	0.012	b_1 [mm]	76.50	B_{ini} [mm]	-30,220
fp	0.012	b_2 [mm]	60.00	C_{ini} [mm]	-20,188
μ	1.4	L_p [mm]	166.00	θ_{4ini} [°]	35
α_{front} [°]	26.5	e [mm]	19	θ_{5ini} [°]	-6
D_{lfront} [mm]	7.0	θ_7 [°]	165	ΔD_{lrear} [mm]	329.93
K_{front} [N/mm]	12.6	L_{pin} [mm]	120		
D_{lrear} [mm]	25.0	h_{pin} [mm]	18.7		
K_{rear} [N/mm]	105.0	r_d [mm]	95		
p_c [mm]	15.875	χ [°]	75.00		
z_p [mm]	15	d_1 [mm]	14.00		
z_c [mm]	39	d_2 [mm]	85.00		
a_{ini} [mm]	94.50	d_3 [mm]	123.00		
b_{ini} [mm]	14.00	d_4 [mm]	1.00		
		d_5 [mm]	12.50		

Fig. 3.1 First set of data necessary for calculation (rider and passenger)

(without parenthesis) and the sole rider loading condition (between parenthesis). The same data joint to some preliminary calculations of the initial parameters have been inserted in the calc sheet whose picture is inserted in Fig. 3.1 (rider and passenger) and in Fig. 3.2 (only rider).

Starting from these initial data it is possible to calculate the loads and stresses depending on the roll angle α and on the speed v. By imposing the speed value from 20 to 200 km/h and by modifying the roll angle from $-54°$ (left bend) to $+54°$ (right bend), the formulae presented in the Chaps. 1 and 2 have been inserted into some calc

W_{max} [W]	93,200	L_r [mm]	565	Fp [N]	3,188
m_{tot} [kg]	325	θ_{1ini} [°]	9	$atan(\mu)$ [°]	54
$JB_{ini} = JB_0$ [mm]	1,440	Xr [mm]	182.00	Φ_p [mm]	75.798
X_{Grear_ini} [mm]	633	Yr [mm]	87.50	Φ_c [mm]	197.074
Y_{G_ini} [mm]	621	Xsr [mm]	216.50	Δ [°]	0.483
R_{W_front} [mm]	300	Ysr [mm]	145.00	β_{ini} [°]	9.483
r_{T_front} [mm]	57	α_{ini} [°]	29.00	θ_{2ini} [°]	35
R_{W_rear} [mm]	310	Xwini [mm]	39.00	θ_{3ini} [°]	25
r_{T_rear} [mm]	115	Ywini [mm]	163.00	A_{ini} [mm]	8,209
fa	0.012	b_1 [mm]	76.50	B_{ini} [mm]	-30,220
fp	0.012	b_2 [mm]	60.00	C_{ini} [mm]	-20,188
μ	1.4	L_p [mm]	166.00	θ_{4ini} [°]	35
α_{front} [°]	26.5	e [mm]	19	θ_{5ini} [°]	-6
D_{ifront} [mm]	7.0	θ_7 [°]	165	ΔD_{irear} [mm]	329.93
K_{front} [N/mm]	12.6	L_{pin} [mm]	120		
D_{irear} [mm]	10.0	h_{pin} [mm]	18.7		
K_{rear} [N/mm]	105.0	r_d [mm]	95		
p_c [mm]	15.875	χ [°]	75.00		
z_p [mm]	15	d_1 [mm]	14.00		
z_c [mm]	39	d_2 [mm]	85.00		
a_{ini} [mm]	94.50	d_3 [mm]	123.00		
b_{ini} [mm]	14.00	d_4 [mm]	1.00		
		d_5 [mm]	12.50		

Fig. 3.2 First set of data necessary for calculation (only rider)

sheets. Each sheet is referred, for a stated load condition (rider plus passenger or the sole rider), to a certain speed value; for example in Fig. 3.3 are reported the results for rider plus passenger with a motorbike ridden at the speed value of 100 km/h. The last column of Fig. 3.3 with the text highlighted in bold style, contains the results for the braking condition with no traction force applied and the motorbike ridden into a straight plane ($\alpha = 0°$); the other data are referred to different roll angles with the maximum traction force applied and without the braking force.

As well indicated by a zoomed view of Fig. 3.3 reported in Fig. 3.4, the diving or the rising of the front suspension (D_{front_calc}) is automatically calculated by the calc sheet, whereas the rotation equilibrium has to be imposed at the null value by changing the rotation angle (β_{act} highlighted in grey in Fig. 3.4). This is a typical recursive problem to be solved with the help of a special *goal search* tool embedded in the calc sheet which is effective for finding the actual angle β_{act}.

At the bottom of the sheet of Fig. 3.3, zoomed and reported in Fig. 3.5, loads and stresses are calculated for different roll angles referred to the symbols presented in the Chap. 2. The loads R_{txx} collected in a unique sheet for the rider plus passenger load condition, for different speeds and for different roll angles, are reported into the picture (tables and diagrams) of Figs. 3.6, 3.7, 3.8 and 3.9.

	20	15	10	5	0	-5	-10	-15	-20	0
α [°]	20	15	10	5	0	-5	-10	-15	-20	0
v [km/h]	100	100	100	100	100	100	100	100	100	100
$1/r_{curve}$ [m⁻¹]	0.0046	0.0034	0.0022	0.0011	0.0000	-0.0011	-0.0022	-0.0034	-0.0046	0.0000
W_{mass} [W]	91,522	88,530	86,488	85,300	84,910	85,300	86,488	88,530	91,522	148,148
W_{mass} [W]	156,094	158,641	160,327	161,290	161,603	161,290	160,327	158,641	156,094	121,109
W [W]	93,200	93,200	93,200	93,200	93,200	93,200	93,200	93,200	93,200	93,200
W_{axle} [W]	91,522	88,530	86,488	85,300	84,910	85,300	86,488	88,530	91,522	93,200
$r_{c,front}$ [mm]	296.6	298.1	299.1	299.8	300.0	299.8	299.1	298.1	296.6	300.0
$r_{c,rear}$ [mm]	303.1	306.1	308.3	309.6	310.0	309.6	308.3	306.1	303.1	310.0
Roll position 1										
$JB=J'Bc=J'B'$ [mm]	1,440.0	1,440.0	1,440.0	1,440.0	1,440.0	1,440.0	1,440.0	1,440.0	1,440.0	1,440.0
$BG_o=BG'=BG''$ [mm]	648.8	648.8	648.8	648.8	648.8	648.8	648.8	648.8	648.8	648.8
ε_{o1} [°]	0.1	0.1	0.0	0.0	0.0	0.0	0.0	0.1	0.1	0.0
μ_{o1} [°]	39.9	39.9	40.0	40.0	40.0	40.0	40.0	39.9	39.9	40.0
$J'B_c$ [mm]	1,440.0	1,440.0	1,440.0	1,440.0	1,440.0	1,440.0	1,440.0	1,440.0	1,440.0	1,440.0
X'_{drear} [mm]	498.0	497.6	497.3	497.1	497.0	497.1	497.3	497.6	498.0	497.0
Y'_o [mm]	725.8	726.3	726.7	726.9	727.0	726.9	726.7	726.3	725.8	727.0
Roll position 2										
$J'G''=JG_o=J'G''$ [mm]	1,035.2	1,035.2	1,035.2	1,035.2	1,035.2	1,035.2	1,035.2	1,035.2	1,035.2	1,035.2
ε_{o2} [°]	0.3	0.2	0.1	0.0	0.0	0.0	0.1	0.2	0.3	0.0
μ_{o2} [°]	24.2	24.3	24.3	24.4	24.4	24.4	24.3	24.3	24.2	24.4
$J'B_o$ [mm]	1,440.0	1,440.0	1,440.0	1,440.0	1,440.0	1,440.0	1,440.0	1,440.0	1,440.0	1,440.0
X_{orear0} [mm]	496.0	496.4	496.7	496.9	497.0	496.9	496.7	496.4	496.0	497.0
Y_{oo} [mm]	721.3	723.8	725.6	726.6	727.0	726.6	725.6	723.8	721.3	727.0
Front suspension										
$D_{front,axle}$ [mm]	-7.00	-7.00	-7.00	-7.00	0.00	0.00	-7.00	-7.00	-7.00	85.28
D_{front} [mm]	0.00	0.00	0.00	0.00	0.00	0.00	0.00	0.00	0.00	85.28
λ [°]	-0.3	-0.3	-0.4	-0.4	-0.4	-0.4	-0.4	-0.3	-0.3	-0.4
J_sB' [mm]	1,440.0	1,440.0	1,440.0	1,440.0	1,440.0	1,440.0	1,440.0	1,440.0	1,440.0	1,403.5
λ' [°]	-0.3	-0.3	-0.4	-0.4	-0.4	-0.4	-0.4	-0.3	-0.3	-0.4
ε_1 [°]	0.0	0.0	0.0	0.0	0.0	0.0	0.0	0.0	0.0	3.1
μ_1 [°]	40.1	40.1	40.0	40.0	40.0	40.0	40.0	40.1	40.1	36.9
$J_sB'_o$ [mm]	1,440.0	1,440.0	1,440.0	1,440.0	1,440.0	1,440.0	1,440.0	1,440.0	1,440.0	1,403.5
X'_{orear} [mm]	496.0	496.4	496.7	496.9	497.0	496.9	496.7	496.4	496.0	518.9
$Y'_{oo}0$ [mm]	721.3	723.8	725.6	726.6	727.0	726.6	725.6	723.8	721.3	699.4
Rear suspension										
β_{ant} [°]	2.9704	3.1933	3.3444	3.43203	3.46073	3.4320	3.3444	3.1933	2.9704	4.2521
Rotation equilibrium	-0	-0	-0	-0	-0	-0	-0	-0	-0	0
AJ_s [mm]	871.1	871.1	871.1	871.1	871.1	871.1	871.1	871.1	871.1	827.1
φ [°]	6.6	6.6	6.6	6.6	6.6	6.6	6.6	6.6	6.6	4.8
ε_1 [°]	4.4	4.3	4.1	4.0	4.0	4.0	4.1	4.3	4.4	1.5
ε' [°]	-4.6	-4.3	-4.2	-4.1	-4.0	-4.1	-4.2	-4.3	-4.6	1.6
μ'_s [°]	19.8	20.0	20.2	20.3	20.3	20.3	20.2	20.0	19.8	22.8
$B'B''$ [mm]	12.9	12.6	12.4	12.2	12.3	12.2	12.4	12.6	12.9	3.7
$G''J_s$ [mm]	1,035.2	1,035.2	1,035.2	1,035.2	1,035.2	1,035.2	1,035.2	1,035.2	1,035.2	970.5
$J_sB'_o$ [mm]	1,452.9	1,452.6	1,452.4	1,452.3	1,452.2	1,452.3	1,452.4	1,452.6	1,452.9	1,407.1
X_{orear} [mm]	478.8	480.1	480.9	481.4	481.6	481.4	480.9	480.1	478.8	512.5
X_{ofront} [mm]	974.1	972.6	971.5	970.8	970.6	970.8	971.5	972.6	974.1	894.6
Y_o [mm]	646.9	652.6	656.6	659.0	659.6	659.0	656.6	652.6	646.9	676.2
AB [mm]	583.4	583.4	583.4	583.4	583.4	583.4	583.4	583.4	583.4	583.4
a [mm]	93.1	93.2	93.2	93.3	93.3	93.3	93.2	93.2	93.1	94.9
b [mm]	21.5	21.1	20.8	20.6	20.6	20.6	20.8	21.1	21.5	11.3
θ [°]	4.36	4.54	4.64	4.70	4.72	4.70	4.64	4.54	4.36	4.61
η [°]	5.13	5.13	5.13	5.13	5.13	5.13	5.13	5.13	5.13	5.12
Xw [mm]	51.89	51.20	50.71	50.43	50.34	50.43	50.71	51.20	51.89	34.36
Yw [mm]	159.37	159.59	159.74	159.83	159.86	159.83	159.74	159.59	159.37	164.04
θ_1 [°]	2.49	2.71	2.86	2.95	2.98	2.95	2.86	2.71	2.49	3.77
θ_2 [°]	28.16	28.39	28.54	28.63	28.66	28.63	28.54	28.39	28.16	29.45
θ_3 [°]	31.32	31.10	30.95	30.86	30.83	30.86	30.95	31.10	31.32	30.04
A [mm²]	3,856.39	3,823.38	3,937.07	4,003.16	4,024.85	4,003.16	3,937.07	3,823.38	3,856.39	5,136.83
B [mm²]	-26,116.78	-26,303.37	-26,431.45	-26,506.34	-26,530.98	-26,506.34	-26,431.45	-26,303.37	-26,116.78	-29,254.75
C [mm²]	-8,005.14	-8,476.33	-8,602.56	-8,694.33	-9,057.91	-8,694.33	-8,602.56	-8,476.33	-8,005.14	-19,984.66
θ_4 [°]	64.36	63.13	62.30	61.81	61.65	61.81	62.30	63.13	64.36	47.46
θ_5 [°]	-35.36	-34.13	-33.30	-32.81	-32.65	-32.81	-33.30	-34.13	-35.36	-16.48
θ_6 [°]	34.02	34.27	34.32	34.34	34.32	34.34	34.32	34.27	34.02	32.52
θ_5 [°]	51.48	51.56	51.62	51.65	51.66	51.65	51.62	51.56	51.48	50.92
Drear [mm]	48.81	46.61	45.10	44.23	43.94	44.23	45.10	46.61	48.81	15.22
F_{rear} [N]	7,760	7,519	7,361	7,269	7,238	7,269	7,361	7,519	7,760	4,223
F'_{rear} [N]	6,134	5,929	5,788	5,707	5,680	5,707	5,788	5,929	6,134	3,145
F_1 [N]	3,295	3,187	3,114	3,071	3,057	3,071	3,114	3,187	3,295	0
F_{chain} [N]	10,134	9,900	9,740	9,647	9,617	9,647	9,740	9,900	10,134	0
F_{hmax} [N]	0	0	0	0	0	0	0	0	0	2,161
Fpa [N]	0	0	0	0	0	0	0	0	0	2,863
Fpp [N]	4,120	4,120	4,120	4,120	4,120	4,120	4,120	4,120	4,120	1,557
$F_{pp}+F_{pa}$ [N]	4,120	4,120	4,120	4,120	4,120	4,120	4,120	4,120	4,120	4,120
F_s [N]	1,500	1,104	727	360	0	-360	-727	-1,104	-1,500	0
Fca [N]	0	0	0	0	0	0	0	0	0	0
Fcagrip [N]	0	0	0	0	0	0	0	0	0	3,588
Feb [N]	1,500	1,104	727	360	0	-360	-727	-1,104	-1,500	7,053
F_s [N]	0	0	0	0	0	0	0	0	0	0
BC [mm]	0.00	0.00	0.00	0.00	0.00	0.00	0.00	0.00	0.00	138.44
θ_{10} [°]	0.00	0.00	0.00	0.00	0.00	0.00	0.00	0.00	0.00	2.04
θ_{11} [°]	0.00	0.00	0.00	0.00	0.00	0.00	0.00	0.00	0.00	9.27
θ_{12} [°]	0.00	0.00	0.00	0.00	0.00	0.00	0.00	0.00	0.00	7.76
θ_{13} [°]	0.00	0.00	0.00	0.00	0.00	0.00	0.00	0.00	0.00	5.78
F_{sb} [N]	0	0	0	0	0	0	0	0	0	4,884
R_{s11} [N]	0	0	0	0	0	0	0	0	0	7,578
R_{s12} [N]	0	0	0	0	0	0	0	0	0	3,727
R_{s21} [N]	0	0	0	0	0	0	0	0	0	1,341
R_{s22} [N]	0	0	0	0	0	0	0	0	0	225
R_{sp1} [N]	0	0	0	0	0	0	0	0	0	7,705
R_{sn1} [N]	0	0	0	0	0	0	0	0	0	3,455
R_{sp2} [N]	0	0	0	0	0	0	0	0	0	1,232
R_{sn2} [N]	0	0	0	0	0	0	0	0	0	269
CK_{rear} [mm]	39	30	20	10	0	-10	-20	-30	-39	2,610
R_{11} [N]	3,336	2,857	2,400	1,973	1,547	1,120	683	227	-260	1,864
R_{12} [N]	2,892	3,374	3,877	4,408	4,969	5,574	6,230	6,950	7,750	1,864
R_{13} [N]	0	0	0	0	0	0	0	0	0	0
R_{21} [N]	16,714	15,894	15,210	14,641	14,170	13,788	13,487	13,264	13,118	430
R_{22} [N]	1,359	789	224	-343	-918	-1,511	-2,130	-2,788	-3,499	-307
R_{23} [N]	3,307	2,695	2,210	1,743	1,286	831	371	-102	-596	10,187
R_{12} [N]	3,034	3,505	3,992	4,502	5,043	5,624	6,256	6,953	7,731	5,487
R_{13} [N]	0	0	0	0	0	0	0	0	0	0
R_{2n} [N]	16,757	15,914	15,203	14,604	14,103	13,692	13,364	13,117	12,954	1,641
R_{22} [N]	633	36	-536	-1,096	-1,653	-2,218	-2,801	-3,412	-4,065	-66
R_1 [N]	3,666	3,541	3,456	3,406	3,390	3,406	3,456	3,541	3,666	5,421
R_2 [N]	13,550	13,219	12,993	12,861	12,817	12,861	12,993	13,219	13,550	-8,546
M_f [Nmm]	708,730	660,593	618,141	580,321	546,333	515,570	487,573	462,017	438,706	419,892
M_t [Nmm]	85,244	123,150	160,746	198,714	237,707	276,396	321,524	367,955	418,754	197,127

Fig. 3.3 Example of calculation for a motorbike speed value of 100 km/h

α [°]	20	15	10	5	0	-5	-10	-15	-20	0
v [km/h]	100	100	100	100	100	100	100	100	100	100
Front suspension										
$D_{front\ calc}$ [mm]	-7.00	-7.00	-7.00	-7.00	-7.00	-7.00	-7.00	-7.00	-7.00	85.28
D_{front} [mm]	0.00	0.00	0.00	0.00	0.00	0.00	0.00	0.00	0.00	85.28
λ [°]	-0.3	-0.3	-0.4	-0.4	-0.4	-0.4	-0.4	-0.3	-0.3	-0.4
J_1B' [mm]	1,440.0	1,440.0	1,440.0	1,440.0	1,440.0	1,440.0	1,440.0	1,440.0	1,440.0	1,403.5
λ' [°]	-0.3	-0.3	-0.4	-0.4	-0.4	-0.4	-0.4	-0.3	-0.3	-0.4
ε_1 [°]	0.0	0.0	0.0	0.0	0.0	0.0	0.0	0.0	0.0	3.1
μ_1 [°]	40.1	40.1	40.0	40.0	40.0	40.0	40.0	40.1	40.1	36.9
$J_1B'_0$ [mm]	1,440.0	1,440.0	1,440.0	1,440.0	1,440.0	1,440.0	1,440.0	1,440.0	1,440.0	1,403.5
X'_0rear [mm]	496.0	496.4	496.7	496.9	497.0	496.9	496.7	496.4	496.0	518.9
$Y'_0 0$ [mm]	721.3	723.8	725.6	726.6	727.0	726.6	725.6	723.8	721.3	699.4
Rear suspension										
β_{act} [°]	2.9704	3.1933	3.3444	3.43203	3.46073	3.4320	3.3444	3.1933	2.9704	4.2521
Rotation equilibrium	-0	-0	-0	-0	-0	-0	-0	-0	-0	0

Fig. 3.4   First enlargement of Fig. 3.3

α [°]	20	15	10	5	0	-5	-10	-15	-20	0
v [km/h]	100	100	100	100	100	100	100	100	100	100
R_{t11} [N]	3,207	2,695	2,210	1,743	1,286	831	371	-102	-596	10,187
R_{t12} [N]	3,034	3,505	3,992	4,502	5,043	5,624	6,256	6,953	7,731	5,487
R_{t13} [N]	0	0	0	0	0	0	0	0	0	0
R_{t21} [N]	16,757	15,914	15,203	14,604	14,103	13,692	13,364	13,117	12,954	1,641
R_{t22} [N]	633	36	-536	-1,096	-1,653	-2,218	-2,801	-3,432	-4,065	-66
R_1 [N]	3,666	3,541	3,456	3,406	3,390	3,406	3,456	3,541	3,666	5,421
R_2 [N]	13,550	13,219	12,993	12,861	12,817	12,861	12,993	13,219	13,550	-8,546
M_f [Nmm]	708,730	660,593	618,141	580,321	546,333	515,570	487,573	462,017	438,706	419,892
M_t [Nmm]	85,244	123,150	160,746	198,714	237,707	278,396	321,524	367,955	418,754	197,127

Fig. 3.5   Second enlargement of Fig. 3.3

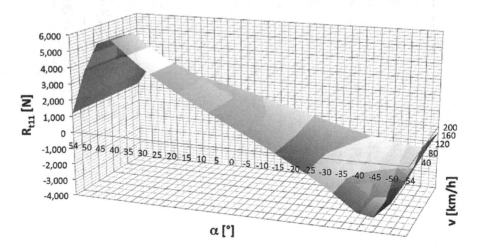

Fig. 3.6   Stress values and diagram of R_{t11} for rider and passenger

Roll angle α [°]	54	50	45	40	35	30	25	20	15	10	5	0	-5	-10	-15	-20	-25	-30	-35	-40	-45	-50	-54	0
200	-512	-102	363	800	1,221	1,633	2,044	2,458	2,879	3,311	3,759	4,226	4,717	5,239	5,799	6,406	7,074	7,820	8,666	9,645	10,806	12,218	12,514	5,487
180	-512	-95	380	826	1,257	1,680	2,101	2,525	2,957	3,401	3,859	4,337	4,840	5,372	5,943	6,561	7,239	7,994	8,850	9,840	11,010	12,433	12,514	5,487
160	-512	-85	401	859	1,302	1,737	2,171	2,609	3,055	3,512	3,984	4,476	4,992	5,538	6,122	6,752	7,443	8,211	9,079	10,081	11,264	12,701	12,514	5,487
140	-512	-74	427	901	1,359	1,811	2,261	2,716	3,179	3,653	4,143	4,652	5,186	5,749	6,350	6,997	7,704	8,488	9,372	10,390	11,590	13,044	12,514	5,487
120	-512	-68	461	955	1,434	1,907	2,380	2,857	3,343	3,841	4,355	4,888	5,444	6,030	6,652	7,321	8,049	8,855	9,760	10,799	12,021	13,220	12,514	5,487
100	-512	-68	507	1,029	1,537	2,040	2,545	3,034	3,505	3,992	4,502	5,043	5,624	6,256	6,953	7,731	8,537	9,367	10,297	11,367	12,620	13,220	12,514	5,487
80	-512	-68	545	1,135	1,635	2,108	2,571	3,034	3,505	3,992	4,502	5,043	5,624	6,256	6,953	7,731	8,613	9,631	10,826	12,225	13,154	13,220	12,514	5,487
60	-512	-68	545	1,140	1,635	2,108	2,571	3,034	3,505	3,992	4,502	5,043	5,624	6,256	6,953	7,731	8,613	9,631	10,826	12,263	13,154	13,220	12,514	5,487
40	-512	-68	545	1,140	1,635	2,108	2,571	3,034	3,505	3,992	4,502	5,043	5,624	6,256	6,953	7,731	8,613	9,631	10,826	12,263	13,154	13,220	12,514	5,487
20	-512	-68	545	1,140	1,635	2,108	2,571	3,034	3,505	3,992	4,502	5,043	5,624	6,256	6,953	7,731	8,613	9,631	10,826	12,263	13,154	13,220	12,514	5,487

(Speed v [km/h] is given in the first column.)

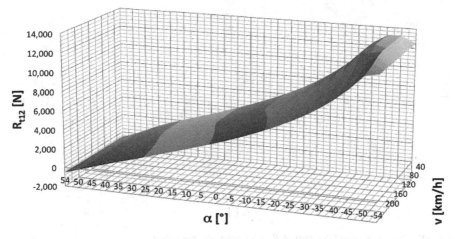

Fig. 3.7 Stress values and diagram of R_{t12} for rider and passenger

Roll angle α [°]	54	50	45	40	35	30	25	20	15	10	5	0	-5	-10	-15	-20	-25	-30	-35	-40	-45	-50	-54	0
200	3,858	8,662	8,763	8,833	8,827	8,775	8,681	8,551	8,388	8,195	7,974	7,727	7,459	7,171	6,867	6,551	6,226	5,900	5,577	5,269	4,990	4,761	2,282	1,641
180	3,858	9,633	9,760	9,812	9,804	9,745	9,641	9,498	9,318	9,104	8,860	8,589	8,292	7,974	7,637	7,287	6,927	6,564	6,205	5,860	5,543	5,278	2,282	1,641
160	3,858	10,846	10,981	11,035	11,024	10,957	10,841	10,681	10,480	10,242	9,969	9,665	9,333	8,977	8,600	8,207	7,803	7,394	6,989	6,598	6,235	5,926	2,282	1,641
140	3,858	12,405	12,550	12,607	12,592	12,516	12,384	12,202	11,974	11,704	11,394	11,049	10,672	10,267	9,838	9,390	8,929	8,463	7,998	7,547	7,126	6,760	2,282	1,641
120	3,858	13,210	14,642	14,703	14,684	14,594	14,441	14,231	13,966	13,653	13,295	12,896	12,460	11,989	11,490	10,968	10,432	9,887	9,344	8,814	8,315	7,191	2,282	1,641
100	3,858	13,210	17,570	17,637	17,611	17,498	17,313	16,757	15,914	15,203	14,604	14,103	13,692	13,364	13,117	12,954	12,550	11,892	11,229	10,589	9,981	7,191	2,282	1,641
80	3,858	13,210	20,201	22,017	20,417	18,962	17,790	16,737	15,914	15,203	14,604	14,103	13,692	13,364	13,117	12,954	12,862	12,914	13,075	13,285	11,482	7,191	2,282	1,641
60	3,858	13,210	20,201	22,308	20,417	18,962	17,760	16,757	15,914	15,203	14,604	14,103	13,692	13,364	13,117	12,954	12,882	12,914	13,075	13,403	11,482	7,191	2,282	1,641
40	3,854	13,210	20,201	22,206	20,417	18,962	17,760	16,757	15,914	15,203	14,604	14,103	13,692	13,364	13,117	12,954	12,882	12,914	13,075	13,403	11,482	7,191	2,282	1,641
20	3,858	13,210	20,201	22,206	20,417	18,962	17,760	16,757	15,914	15,203	14,604	14,103	13,692	13,364	13,117	12,954	12,882	12,914	13,075	13,403	11,482	7,191	2,282	1,641

(Speed v [km/h] is given in the first column.)

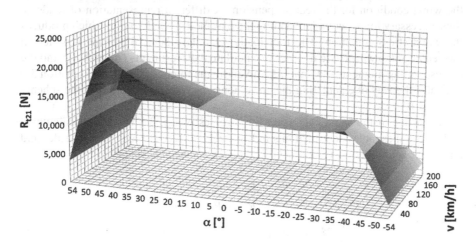

Fig. 3.8 Stress values and diagram of R_{t21} for rider and passenger

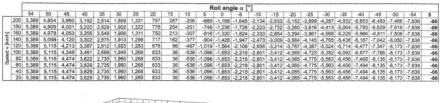

		Roll angle α [°]																						
		54	50	45	40	35	30	25	20	15	10	5	0	-5	-10	-15	-20	-25	-30	-35	-40	-45	-50	-54	0
Speed v [km/h]	200	5,389	4,854	3,960	3,192	2,514	1,899	1,331	797	287	-206	-689	-1,168	-1,648	-2,134	-2,633	-3,152	-3,699	-4,287	-4,932	-5,653	-6,483	-7,466	-7,638	-66
	180	5,389	4,909	4,001	3,220	2,529	1,902	1,322	778	254	-251	-746	-1,236	-1,726	-2,223	-2,732	-3,260	-3,816	-4,413	-5,064	-5,793	-6,529	-7,619	-7,638	-66
	160	5,389	4,978	4,053	3,255	2,548	1,906	1,311	750	213	-307	-816	-1,320	-1,824	-2,333	-2,854	-3,394	-3,961	-4,568	-5,229	-5,966	-6,811	-7,808	-7,638	-66
	140	5,389	5,068	4,120	3,302	2,575	1,913	1,298	717	162	-377	-904	-1,426	-1,947	-2,473	-3,009	-3,564	-4,145	-4,765	-5,438	-6,187	-7,042	-8,050	-7,638	-66
	120	5,389	5,115	4,213	3,367	2,612	1,923	1,283	676	96	-467	-1,019	-1,564	-2,108	-2,656	-3,214	-3,787	-4,387	-5,024	-5,714	-6,477	-7,347	-8,173	-7,638	-66
	100	5,389	5,115	4,348	3,461	2,669	1,945	1,269	633	36	-536	-1,096	-1,653	-2,218	-2,801	-3,412	-4,065	-4,723	-5,382	-6,092	-6,877	-7,766	-8,173	-7,638	-66
	80	5,389	5,115	4,474	3,622	2,735	1,960	1,268	633	36	-536	-1,096	-1,653	-2,218	-2,801	-3,412	-4,065	-4,775	-5,563	-6,456	-7,468	-8,135	-8,173	-7,638	-66
	60	5,389	5,115	4,474	3,629	2,735	1,960	1,268	633	36	-536	-1,096	-1,653	-2,218	-2,801	-3,412	-4,065	-4,775	-5,563	-6,456	-7,494	-8,135	-8,173	-7,638	-66
	40	5,389	5,115	4,474	3,629	2,735	1,960	1,268	633	36	-536	-1,096	-1,653	-2,218	-2,801	-3,412	-4,065	-4,775	-5,563	-6,456	-7,494	-8,135	-8,173	-7,638	-66
	20	5,389	5,115	4,474	3,629	2,735	1,960	1,268	633	36	-536	-1,096	-1,653	-2,218	-2,801	-3,412	-4,065	-4,775	-5,563	-6,456	-7,494	-8,135	-8,173	-7,638	-66

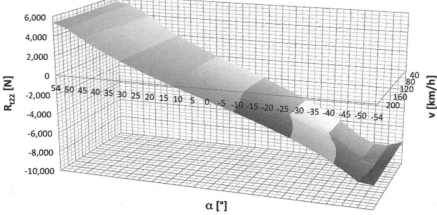

Fig. 3.9 Stress values and diagram of R_{t22} for rider and passenger

3.3 Global Stresses Calculation

The loads trend highlighted in Figs. 3.6, 3.7, 3.8 and 3.9 cannot easily point out the worst condition for the rear suspension. A different representation of loads is, thus, necessary for better understanding the parameters combination which produces the worst stress condition. The four loads R_{txx} have been, therefore, reported in the center of the wheel axle in order to calculate their resultants (R_1 and R_2) and the corresponding moments (M_f and M_t) given by the forces translations. Equation 3.1 indicates the way of calculating the aforementioned stresses referred to the scheme of Fig. 3.10.

By means of the same calc sheet the stresses of Eq. 3.1 have been calculated and reported as tables and diagrams in Figs. 3.11, 3.12, 3.13 and 3.14 for rider and passenger and in Figs. 3.15, 3.16, 3.17 and 3.18 for the sole rider.

$$R_1 = R_{t12} + R_{t22}$$
$$R_2 = R_{t21} - R_{t11}$$
$$M_f = (R_{t11} + R_{t21}) \times \frac{d_2 - d_1}{2}$$
$$M_t = (R_{t12} - R_{t22}) \times \frac{d_2 - d_1}{2}$$

(3.1)

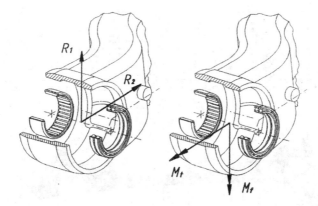

Fig. 3.10 Scheme of stresses R_1, R_2, M_f and M_t

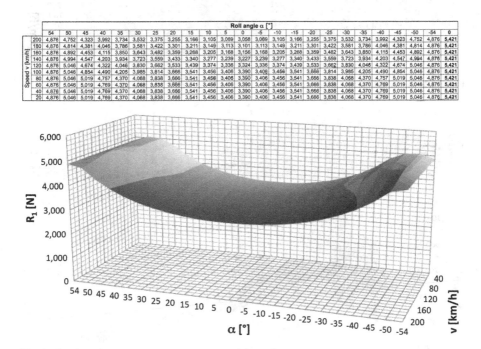

		Roll angle α [°]																						
		54	50	45	40	35	30	25	20	15	10	5	0	-5	-10	-15	-20	-25	-30	-35	-40	-45	-50	-54	0
Speed v [km/h]	200	4,876	4,752	4,323	3,992	3,734	3,532	3,375	3,255	3,166	3,105	3,069	3,058	3,069	3,105	3,166	3,255	3,375	3,532	3,734	3,992	4,323	4,752	4,876	5,421
	180	4,876	4,814	4,381	4,046	3,786	3,581	3,422	3,301	3,211	3,149	3,113	3,101	3,113	3,149	3,211	3,301	3,422	3,581	3,786	4,046	4,381	4,814	4,876	5,421
	160	4,876	4,892	4,453	4,115	3,850	3,643	3,482	3,359	3,268	3,205	3,168	3,156	3,168	3,205	3,268	3,359	3,482	3,643	3,850	4,115	4,453	4,892	4,876	5,421
	140	4,876	4,994	4,547	4,203	3,934	3,723	3,559	3,433	3,340	3,277	3,239	3,227	3,239	3,277	3,340	3,433	3,559	3,723	3,934	4,203	4,547	4,994	4,876	5,421
	120	4,876	5,046	4,674	4,322	4,046	3,830	3,662	3,533	3,439	3,374	3,336	3,324	3,336	3,374	3,439	3,533	3,662	3,830	4,046	4,322	4,674	5,046	4,876	5,421
	100	4,876	5,046	4,854	4,490	4,205	3,985	3,814	3,666	3,541	3,456	3,406	3,390	3,406	3,456	3,541	3,666	3,814	3,985	4,205	4,490	4,854	5,046	4,876	5,421
	80	4,876	5,046	5,019	4,757	4,370	4,068	3,838	3,666	3,541	3,456	3,406	3,390	3,406	3,456	3,541	3,666	3,838	4,068	4,370	4,757	5,019	5,046	4,876	5,421
	60	4,876	5,046	5,019	4,769	4,370	4,068	3,838	3,666	3,541	3,456	3,406	3,390	3,406	3,456	3,541	3,666	3,838	4,068	4,370	4,769	5,019	5,046	4,876	5,421
	40	4,876	5,046	5,019	4,769	4,370	4,068	3,838	3,666	3,541	3,456	3,406	3,390	3,406	3,456	3,541	3,666	3,838	4,068	4,370	4,769	5,019	5,046	4,876	5,421
	20	4,876	5,046	5,019	4,769	4,370	4,068	3,838	3,666	3,541	3,456	3,406	3,390	3,406	3,456	3,541	3,666	3,838	4,068	4,370	4,769	5,019	5,046	4,876	5,421

Fig. 3.11 Stress values and diagram of R_1 for rider and passenger

		Roll angle α [°]																					
		54	50	45	40	35	30	25	20	15	10	5	0	-5	-10	-15	-20	-25	-30	-35	-40	-45	-50	-54	0
Speed v [km/h]	200	2,678	6,146	6,355	6,524	6,665	6,782	6,879	6,956	7,016	7,058	7,083	7,091	7,083	7,058	7,016	6,956	6,879	6,782	6,665	6,524	6,355	6,146	2,678	-8,546
	180	2,678	6,836	7,062	7,246	7,399	7,527	7,633	7,718	7,784	7,831	7,859	7,868	7,859	7,831	7,784	7,718	7,633	7,527	7,399	7,246	7,062	6,836	2,678	-8,546
	160	2,678	7,700	7,945	8,147	8,317	8,459	8,576	8,671	8,745	8,797	8,828	8,838	8,828	8,797	8,745	8,671	8,576	8,459	8,317	8,147	7,945	7,700	2,678	-8,546
	140	2,678	8,809	9,081	9,306	9,496	9,656	9,789	9,897	9,980	10,039	10,074	10,086	10,074	10,039	9,980	9,897	9,789	9,656	9,496	9,306	9,081	8,809	2,678	-8,546
	120	2,678	9,381	10,594	10,851	11,069	11,253	11,406	11,530	11,624	11,689	11,729	11,742	11,729	11,689	11,624	11,530	11,406	11,253	11,069	10,851	10,594	9,381	2,678	-8,546
	100	2,678	9,381	12,712	13,013	13,269	13,478	13,652	13,550	13,219	12,993	12,861	12,817	12,861	12,993	13,219	13,550	13,652	13,478	13,269	13,013	12,712	9,381	2,678	-8,546
	80	2,678	9,381	14,614	16,229	15,367	14,595	14,001	13,550	13,219	12,993	12,861	12,817	12,861	12,993	13,219	13,550	14,001	14,595	15,367	16,229	14,614	9,381	2,678	-8,546
	60	2,678	9,381	14,614	16,368	15,367	14,595	14,001	13,550	13,219	12,993	12,861	12,817	12,861	12,993	13,219	13,550	14,001	14,595	15,367	16,368	14,614	9,381	2,678	-8,546
	40	2,675	9,381	14,614	16,368	15,367	14,595	14,001	13,550	13,219	12,993	12,861	12,817	12,861	12,993	13,219	13,550	14,001	14,595	15,367	16,368	14,614	9,381	2,678	-8,546
	20	2,678	9,381	14,614	16,368	15,367	14,595	14,001	13,550	13,219	12,993	12,861	12,817	12,861	12,993	13,219	13,550	14,001	14,595	15,367	16,368	14,614	9,381	2,678	-8,546

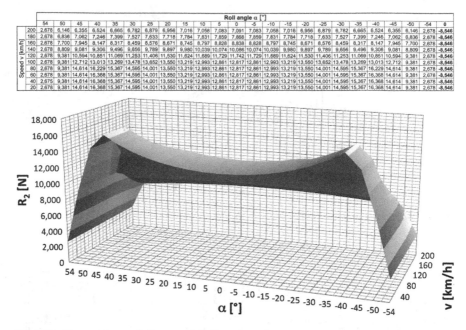

Fig. 3.12 Stress values and diagram of R_2 for rider and passenger

		Roll angle α [°]																					
		54	50	45	40	35	30	25	20	15	10	5	0	-5	-10	-15	-20	-25	-30	-35	-40	-45	-50	-54	0
Speed v [km/h]	200	178,822	396,824	398,027	395,547	390,129	382,232	372,176	360,199	346,502	331,267	314,872	298,897	276,132	258,690	238,507	218,160	197,885	178,105	159,392	142,508	128,677	119,834	66,942	419,892
	180	178,822	441,286	442,286	439,415	433,376	424,644	413,547	400,338	385,232	368,423	350,106	330,476	309,740	288,125	265,901	243,353	220,842	198,816	177,864	158,809	142,582	132,081	66,942	419,892
	160	178,822	496,718	497,804	494,246	487,436	477,660	465,263	450,516	433,647	414,872	394,401	372,451	349,253	325,058	300,150	274,891	249,548	224,718	200,984	179,208	160,668	147,432	66,942	419,892
	140	178,822	568,050	568,720	564,734	556,943	545,826	531,760	515,035	495,902	474,598	451,389	426,426	400,083	372,548	344,190	315,383	286,474	256,044	230,741	205,475	183,550	167,344	66,942	419,892
	120	178,822	604,864	663,524	658,734	649,617	636,718	620,430	601,070	578,975	554,410	527,599	498,792	468,253	436,273	403,189	369,406	335,742	302,524	270,474	240,575	214,255	177,507	66,942	419,892
	100	178,822	604,864	796,211	790,301	779,357	763,904	744,598	708,730	660,593	618,141	580,321	546,333	515,570	487,573	482,017	438,706	406,430	365,833	328,227	289,878	257,405	177,507	66,942	419,892
	80	178,822	604,864	915,455	987,047	904,083	828,179	763,954	708,730	660,593	618,141	580,321	546,333	515,570	487,573	462,017	438,706	417,586	398,800	382,788	367,063	298,446	177,507	66,942	419,892
	60	178,822	604,864	915,455	995,561	904,083	828,179	763,954	708,730	660,593	618,141	580,321	546,333	515,570	487,573	462,017	438,706	417,586	398,800	382,788	370,523	298,446	177,507	66,942	419,892
	40	178,628	604,864	915,455	995,561	904,083	828,179	763,954	708,730	660,593	618,141	580,321	546,333	515,570	487,573	462,017	438,706	417,586	398,800	382,788	370,523	298,446	177,507	66,942	419,892
	20	178,822	604,864	915,455	995,561	904,083	828,179	763,954	708,730	660,593	618,141	580,321	546,333	515,570	487,573	462,017	438,706	417,596	398,800	382,788	370,523	298,446	177,507	66,942	419,892

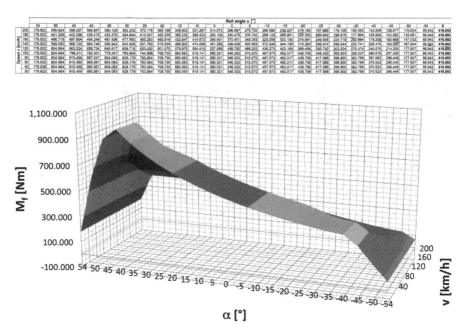

Fig. 3.13 Stress values and diagram of M_f for rider and passenger

	Roll angle α [°]																							
	54	50	45	40	35	30	25	20	15	10	5	0	-5	-10	-15	-20	-25	-30	-35	-40	-45	-50	-54	0
200	-209.484	-175.969	-127.704	-84.930	-45.902	-9.438	25.305	58.955	92.006	124.868	157.909	191.461	225.954	261.739	299.321	339.303	382.458	429.797	482.707	543.108	613.768	698.815	715.377	197.127
180	-209.484	-177.627	-128.546	-84.973	-45.149	-7.887	27.659	62.116	95.975	129.044	163.484	197.846	233.092	269.829	307.940	348.623	392.446	440.433	493.960	554.960	628.210	711.858	715.377	197.127
160	-209.484	-179.742	-129.639	-85.065	-44.248	-5.990	30.557	66.016	100.884	135.555	170.391	205.734	241.942	279.416	318.633	360.190	404.852	453.840	507.939	569.688	641.678	728.081	715.377	197.127
140	-209.484	-182.532	-131.311	-85.250	-43.196	-3.620	34.208	70.958	107.108	143.063	179.171	215.768	253.204	291.876	332.252	374.925	420.661	470.480	525.771	588.484	661.430	748.813	715.377	197.127
120	-209.484	-184.006	-133.194	-85.614	-41.818	-0.583	38.947	77.401	115.267	152.944	190.764	229.045	268.120	308.364	350.237	394.341	441.503	492.692	549.306	613.310	687.540	759.461	715.377	197.127
100	-209.484	-184.006	-136.354	-86.356	-40.177	3.392	45.298	85.244	123.150	160.746	198.714	237.707	278.396	321.524	367.955	418.754	470.711	523.587	581.812	647.833	723.680	759.461	715.377	197.127
80	-209.484	-184.006	-139.453	-88.260	-39.050	5.238	46.255	85.244	123.150	160.746	198.714	237.707	278.396	321.524	367.955	418.754	475.285	539.367	613.515	699.111	755.789	759.461	715.377	197.127
60	-209.484	-184.006	-139.453	-88.372	-39.050	5.238	46.255	85.244	123.150	160.746	198.714	237.707	278.396	321.524	367.955	418.754	475.285	539.367	613.515	701.368	755.789	759.461	715.377	197.127
40	-209.493	-184.006	-139.453	-88.372	-39.050	5.238	46.255	85.244	123.150	160.746	198.714	237.707	278.396	321.524	367.955	418.754	475.285	539.367	613.515	701.368	755.789	759.461	715.377	197.127
20	-209.493	-184.006	-139.453	-88.372	-39.050	5.238	46.255	85.244	123.150	160.746	198.714	237.707	278.396	321.524	367.955	418.754	475.285	539.367	613.515	701.368	755.789	759.461	715.377	197.127

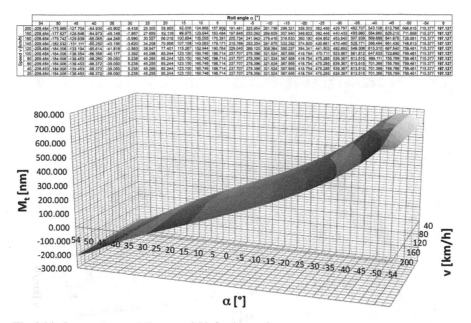

Fig. 3.14 Stress values and diagram of M_t for rider and passenger

	Roll angle α [°]																							
	54	50	45	40	35	30	25	20	15	10	5	0	-5	-10	-15	-20	-25	-30	-35	-40	-45	-50	-54	0
200	3,106	3,067	2,804	2,589	2,420	2,288	2,185	2,106	2,048	2,008	1,985	1,977	1,985	2,008	2,048	2,106	2,185	2,288	2,420	2,589	2,804	3,067	3,106	3,865
180	3,106	3,067	2,840	2,621	2,451	2,317	2,213	2,133	2,074	2,033	2,009	2,002	2,009	2,033	2,074	2,133	2,213	2,317	2,451	2,621	2,840	3,067	3,106	3,865
160	3,106	3,067	2,884	2,662	2,489	2,353	2,247	2,166	2,106	2,064	2,040	2,032	2,040	2,064	2,106	2,166	2,247	2,353	2,489	2,662	2,884	3,067	3,106	3,865
140	3,106	3,067	2,939	2,716	2,538	2,399	2,291	2,208	2,146	2,104	2,079	2,071	2,079	2,104	2,146	2,208	2,291	2,399	2,538	2,716	2,939	3,067	3,106	3,865
120	3,106	3,067	2,939	2,787	2,604	2,461	2,349	2,264	2,200	2,157	2,131	2,123	2,131	2,157	2,200	2,264	2,349	2,461	2,604	2,787	2,939	3,067	3,106	3,865
100	3,106	3,067	2,939	2,823	2,697	2,545	2,431	2,343	2,278	2,234	2,208	2,200	2,208	2,234	2,278	2,343	2,431	2,548	2,697	2,823	2,939	3,067	3,106	3,865
80	3,106	3,067	2,939	2,823	2,729	2,662	2,531	2,412	2,326	2,268	2,234	2,223	2,234	2,268	2,326	2,412	2,531	2,662	2,729	2,823	2,939	3,067	3,106	3,865
60	3,106	3,067	2,939	2,823	2,729	2,662	2,531	2,412	2,326	2,268	2,234	2,223	2,234	2,268	2,326	2,412	2,531	2,662	2,729	2,823	2,939	3,067	3,106	3,865
40	3,106	3,067	2,939	2,823	2,729	2,662	2,531	2,412	2,326	2,268	2,234	2,223	2,234	2,268	2,326	2,412	2,531	2,662	2,729	2,823	2,939	3,067	3,106	3,865
20	3,106	3,067	2,939	2,823	2,729	2,662	2,531	2,412	2,326	2,268	2,234	2,223	2,234	2,268	2,326	2,412	2,531	2,662	2,729	2,823	2,939	3,067	3,106	3,865

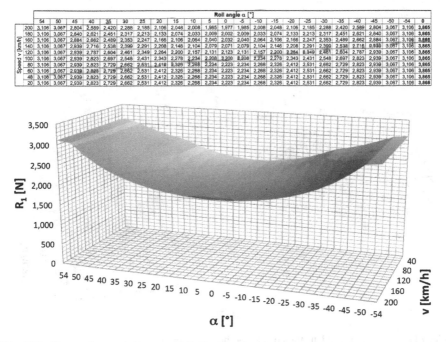

Fig. 3.15 Stress values and diagram of R_1 for the sole rider

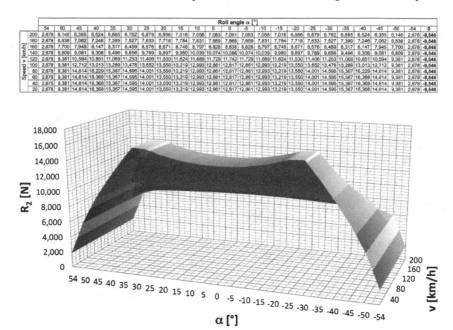

Fig. 3.16 Stress values and diagram of R_2 for the sole rider

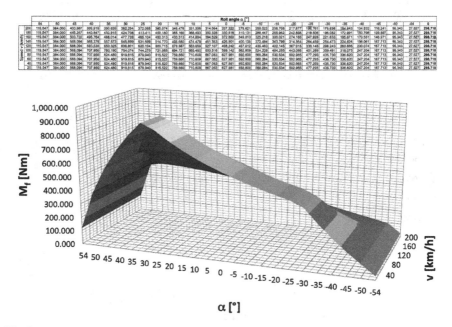

Fig. 3.17 Stress values and diagram of M_f for the sole rider

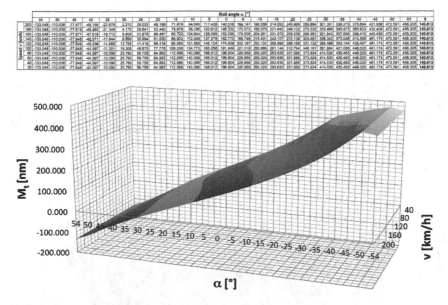

										Roll angle α [°]													
	64	50	45	40	35	30	25	20	15	10	5	0	-5	-10	-15	-20	-25	-30	-35	-40	-45	-50	-54	0
200	-133.048	-110.006	-77.977	-49.155	-22.679	2.210	26.033	49.188	71.976	94.649	117.426	140.518	164.147	188.589	214.052	240.999	269.884	301.351	336.275	375.884	421.938	472.591	458.305	140.813
180	-133.048	-110.006	-77.919	-48.463	-21.349	4.175	28.641	52.442	75.878	99.196	122.612	146.332	170.573	195.579	221.642	249.132	278.532	310.486	345.873	385.923	432.406	472.591	458.305	140.813
160	-133.048	-110.006	-77.871	-47.619	-19.712	6.608	31.876	60.481	80.723	104.844	129.055	153.556	178.559	204.301	231.072	259.239	289.281	321.843	357.806	398.410	445.436	472.591	458.305	140.813
140	-133.048	-110.006	-77.848	-48.571	-17.644	9.699	35.994	61.530	86.903	112.049	137.275	162.773	188.748	215.431	243.107	272.138	303.001	336.342	373.048	414.365	461.174	472.591	458.305	140.813
120	-133.048	-110.006	-77.848	-49.239	-14.950	13.756	41.415	68.418	95.055	121.586	148.124	174.939	202.197	230.124	258.995	289.169	321.122	355.499	393.194	435.467	461.174	472.591	458.305	140.813
100	-133.048	-110.006	-77.848	-44.597	-11.301	19.306	48.870	77.776	108.336	134.772	163.256	191.949	221.019	250.066	281.144	312.794	346.161	381.984	421.056	446.020	461.174	472.591	458.305	140.813
80	-133.048	-110.006	-77.848	-44.597	-10.090	25.789	56.758	84.992	112.586	140.096	168.012	196.804	226.959	259.020	293.630	331.586	373.924	414.530	430.483	446.020	461.174	472.591	458.305	140.813
60	-133.048	-110.006	-77.848	-44.597	-10.090	25.789	56.758	84.992	112.586	140.096	168.012	196.804	226.959	259.020	293.630	331.586	373.924	414.530	430.483	446.020	461.174	472.591	458.305	140.813
40	-133.048	-110.006	-77.848	-44.597	-10.090	25.789	56.758	84.992	112.586	140.096	168.012	196.804	228.959	259.020	293.630	331.586	373.924	414.530	430.483	446.020	461.174	472.591	458.305	140.813
20	-133.048	-110.006	-77.848	-44.597	-10.090	25.789	56.758	84.992	112.588	140.096	168.012	198.804	228.959	259.020	293.630	331.586	373.924	414.535	430.483	446.020	461.174	472.591	458.305	140.813

Fig. 3.18 Stress values and diagram of M_t for the sole rider

3.4 Results Analysis and Comments

The analysis of results plotted from Figs. 3.11, 3.12, 3.13, 3.14, 3.15, 3.16, 3.17, and Fig. 3.18, is useful to understand the stress trends depending on the motion parameters (α and v).

The values assumed by the R_1 stress, which represents the action orthogonal to the longitudinal axis of the rear suspension, are proportional to the mass and centripetal loads (Fp_p and Fc_p). Since the mass load does not change significantly whereas the centripetal load is higher when the absolute value of the angle α is higher, the trend of Figs. 3.11 and 3.15 is well explained. The small decrease at the extreme values of the roll angle is due to the reduction of the tyre grip and, therefore, to the reduction of the load transmittable to the ground; thus there is a progressive reduction of Fp_p and Fc_p balanced by the increase of the Fp_a and Fc_a. This occurrence is confirmed by the diving of the front suspension whose diagram is reported in Fig. 3.19. No significant dependence on motorbike speed is pointed out.

The R_2 stress, which represents the action along the longitudinal axis of the rear suspension, is strictly linked to the traction forces F_{chain} and F_t. For this reason when the traction moment or the traction force decrease also the R_2 value decreases. This situation is well highlighted in correspondence of the external values of the roll angle corresponding to a reduction of the traction force and when the motorbike speed is growing up corresponding to a smooth increase of power and, at the same

		Roll angle α [°]																							
		54	50	45	40	35	30	25	20	15	10	5	0	-5	-10	-15	-20	-25	-30	-35	-40	-45	-50	-54	0
Speed v [km/h]	200	63	42	36	30	26	23	20	18	17	16	15	15	15	16	17	18	20	23	26	30	36	42	63	85
	180	63	40	33	28	23	20	17	15	14	13	12	12	12	13	14	15	17	20	23	28	33	40	63	85
	160	63	36	29	24	20	16	14	12	10	9	8	8	8	9	10	12	14	16	20	24	29	36	63	85
	140	63	32	25	19	15	12	9	7	5	4	4	3	4	4	5	7	9	12	15	19	25	32	63	85
	120	63	29	19	13	9	5	3	0	0	0	0	0	0	0	0	0	0	3	5	9	13	19	29	85
	100	63	29	11	5	1	0	0	0	0	0	0	0	0	0	0	0	0	0	1	5	11	29	63	85
	80	63	29	4	0	0	0	0	0	0	0	0	0	0	0	0	0	0	0	0	0	4	29	63	85
	60	63	29	4	0	0	0	0	0	0	0	0	0	0	0	0	0	0	0	0	0	4	29	63	85
	40	63	29	4	0	0	0	0	0	0	0	0	0	0	0	0	0	0	0	0	0	4	29	63	85
	20	63	29	4	0	0	0	0	0	0	0	0	0	0	0	0	0	0	0	0	0	4	29	63	85

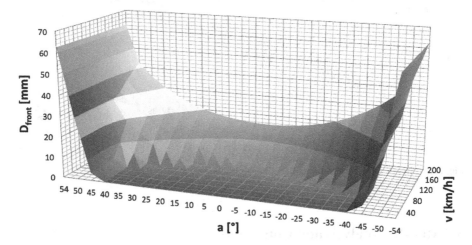

Fig. 3.19 Values and diagram of diving of front suspension D_{front} for rider and passenger

time, to a strong reduction of the traction force given by Eqs. 2.2 and .2.3 of Chap. 2. The same trend of Figs. 3.12 and 3.16 is, therefore, shown in Figs. 3.20 and 3.21.

The M_f trend presented in the diagrams of Figs. 3.13 and 3.17 is due to the arms of the two main loads (F_{chain} and F_t) that produce the M_f moment. When the roll angle increases, the F_{chain} arm remains almost the same whereas the F_t arm increases proportionally to the Y' distance reported in the sketch of Fig. 3.22.

Conversely the arm of torsion loads (F_p and F_c) which produces the M_t increases when the roll angle α decreases because the hub of the rear wheel enlarges its distance from the ground proportionally to the X distance reported in the sketch of Fig. 3.22.

A final comment deserves to be mentioned. All the trends of stresses show different behaviors if the motorbike is driven at low or at high speed. In the case of low values of speed the maximum power transmittable is limited by the capsizing limit so that, in this case, the mass load is, essentially and wholly, applied to the rear wheel and axle, thus, to the rear suspension. When the speed increases it is possible to transfer the total traction power to the rear wheel and, at the same time, to transfer some of the weight forcesto the front wheel. For this reason when the speed value increases the global stresses acting on the rear suspension decrease.

Speed v [km/h]	Roll angle α [°]																							0
	54	50	45	40	35	30	25	20	15	10	5	0	-5	-10	-15	-20	-25	-30	-35	-40	-45	-50	-54	
200	2,079	4,578	4,704	4,820	4,924	5,015	5,094	5,160	5,211	5,248	5,270	5,278	5,270	5,248	5,211	5,160	5,094	5,015	4,924	4,820	4,704	4,578	2,079	0
180	2,079	5,087	5,227	5,355	5,471	5,573	5,660	5,733	5,790	5,831	5,856	5,864	5,856	5,831	5,790	5,733	5,660	5,573	5,471	5,355	5,227	5,087	2,079	0
160	2,079	5,723	5,880	6,025	6,155	6,269	6,368	6,450	6,514	6,560	6,588	6,597	6,588	6,560	6,514	6,450	6,368	6,269	6,155	6,025	5,880	5,723	2,079	0
140	2,079	6,541	6,720	6,885	7,034	7,165	7,278	7,371	7,444	7,497	7,529	7,540	7,529	7,497	7,444	7,371	7,278	7,165	7,034	6,885	6,720	6,541	2,079	0
120	2,079	6,963	7,841	8,033	8,206	8,359	8,491	8,600	8,685	8,747	8,784	8,796	8,784	8,747	8,685	8,600	8,491	8,359	8,206	8,033	7,841	6,963	2,079	0
100	2,079	6,963	9,409	9,639	9,847	10,031	10,189	10,134	9,900	9,740	9,647	9,617	9,647	9,740	9,900	10,134	10,189	10,031	9,847	9,639	9,409	6,963	2,079	0
80	2,079	6,963	10,818	12,049	11,430	10,877	10,453	10,134	9,900	9,740	9,647	9,617	9,647	9,740	9,900	10,134	10,453	10,877	11,430	12,049	10,818	6,963	2,079	0
60	2,079	6,963	10,818	12,154	11,430	10,877	10,453	10,134	9,900	9,740	9,647	9,617	9,647	9,740	9,900	10,134	10,453	10,877	11,430	12,154	10,818	6,963	2,079	0
40	2,076	6,963	10,818	12,154	11,430	10,877	10,453	10,134	9,900	9,740	9,647	9,617	9,647	9,740	9,900	10,134	10,453	10,877	11,430	12,154	10,818	6,963	2,079	0
20	2,079	6,963	10,818	12,154	11,430	10,877	10,453	10,134	9,900	9,740	9,647	9,617	9,647	9,740	9,900	10,134	10,453	10,877	11,430	12,154	10,818	6,963	2,079	0

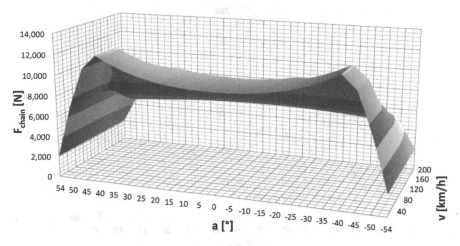

Fig. 3.20 Stress values and diagram of F_{chain} for rider and passenger

Speed v [km/h]	Roll angle α [°]																							0
	54	50	45	40	35	30	25	20	15	10	5	0	-5	-10	-15	-20	-25	-30	-35	-40	-45	-50	-54	
200	780	1,678	1,678	1,678	1,678	1,678	1,678	1,678	1,678	1,678	1,678	1,678	1,678	1,678	1,678	1,678	1,678	1,678	1,678	1,678	1,678	1,678	780	0
180	780	1,864	1,864	1,864	1,864	1,864	1,864	1,864	1,864	1,864	1,864	1,864	1,864	1,864	1,864	1,864	1,864	1,864	1,864	1,864	1,864	1,864	780	0
160	780	2,097	2,097	2,097	2,097	2,097	2,097	2,097	2,097	2,097	2,097	2,097	2,097	2,097	2,097	2,097	2,097	2,097	2,097	2,097	2,097	2,097	780	0
140	780	2,397	2,397	2,397	2,397	2,397	2,397	2,397	2,397	2,397	2,397	2,397	2,397	2,397	2,397	2,397	2,397	2,397	2,397	2,397	2,397	2,397	780	0
120	780	2,551	2,796	2,796	2,796	2,796	2,796	2,796	2,796	2,796	2,796	2,796	2,796	2,796	2,796	2,796	2,796	2,796	2,796	2,796	2,796	2,551	780	0
100	780	2,551	3,355	3,355	3,355	3,355	3,355	3,295	3,187	3,114	3,071	3,057	3,071	3,114	3,187	3,295	3,355	3,355	3,355	3,355	3,355	2,551	780	0
80	780	2,551	3,858	4,194	3,894	3,638	3,442	3,295	3,187	3,114	3,071	3,057	3,071	3,114	3,187	3,295	3,442	3,638	3,894	4,194	3,858	2,551	780	0
60	780	2,551	3,858	4,230	3,894	3,638	3,442	3,295	3,187	3,114	3,071	3,057	3,071	3,114	3,187	3,295	3,442	3,638	3,894	4,230	3,858	2,551	780	0
40	779	2,551	3,858	4,230	3,894	3,638	3,442	3,295	3,187	3,114	3,071	3,057	3,071	3,114	3,187	3,295	3,442	3,638	3,894	4,230	3,858	2,551	780	0
20	780	2,551	3,858	4,230	3,894	3,638	3,442	3,295	3,187	3,114	3,071	3,057	3,071	3,114	3,187	3,295	3,442	3,638	3,894	4,230	3,858	2,551	780	0

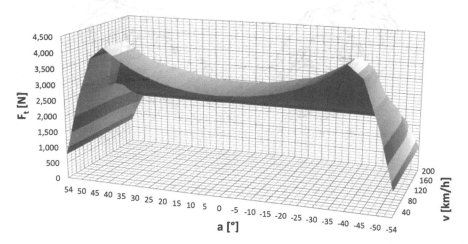

Fig. 3.21 Stress values and diagram of F_t for rider and passenger

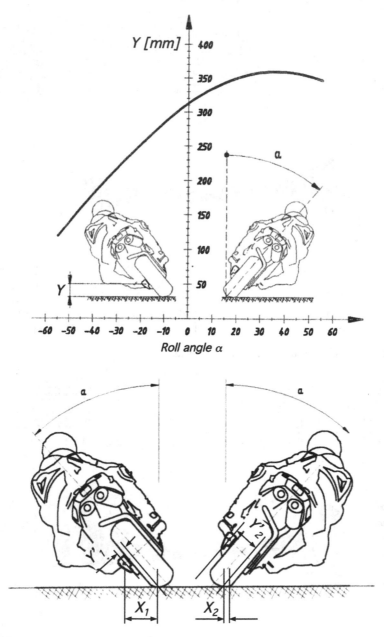

Fig. 3.22 Flection and torsion arms (Y′ and X) depending on the roll angle α

3.5 The Maximum Stresses Combination

The combination of parameters which produces the maximum stress conditions on the rear suspension is difficult to define because, as well indicated by tables and diagrams presented in the previous section, there is no simultaneous presence of the maximum values for R_1, R_2, M_f and M_t. Moreover the two initial load conditions (rider plus passenger and the sole rider) bring to different maximum stress conditions in term of combination of parameters, so that it is impossible to define a unique combination which can be considered as the worst for the rear suspension. Conversely it is necessary to consider at least three different loading conditions for each of the two initial loading cases, thus the tables containing the results in function of the roll angle and of the motorbike speed have been extensively analyzed in order to highlight the most severe stress combinations. In Figs. 3.23 and 3.24 are computed and highlighted through the calc sheet, the maximum stresses depending on the roll angle and on the motorbike speed parameters.

In the case of rider and passenger (see table of Fig. 3.23 joint to the data included from Figs. 3.11, 3.12, 3.13, and 3.14) the three combinations of parameters are the following:

R_1

	54	50	45	40	35	30	25	20	15	10	5	0	-5	-10	-15	-20	-25	-30	-35	-40	-45	-50	-54
200																							4,876
180																							4,876
160		4,892																				4,892	
140		4,994																				4,994	
120		5,046																					
100		5,046																					
80		5,046																				5,046	
60		5,046																					
40																						5,046	
20		5,046																				5,046	

R_2

	54	50	45	40	35	30	25	20	15	10	5	0	-5	-10	-15	-20	-25	-30	-35	-40	-45	-50	-54
200												7,091											
180												7,868											
160												8,838											
140												10,086											
120												11,742											
100						#											#						
80				16,229																16,229			
60				16,368																16,368			
40				16,368																16,368			
20				16,368																16,368			

M_f

	54	50	45	40	35	30	25	20	15	10	5	0	-5	-10	-15	-20	-25	-30	-35	-40	-45	-50	-54
200			398.027																				
180			442.286																				
160			497.604																				
140			568.720																				
120			663.524																				
100			796.211																				
80				987.047																			
60				995.561																			
40				995.561																			
20				995.561																			

M_t

	54	50	45	40	35	30	25	20	15	10	5	0	-5	-10	-15	-20	-25	-30	-35	-40	-45	-50	-54
200																							715
180																							715
160																						728.081	
140																						748.813	
120																						759.461	
100																						759.461	
80																						759.461	
60																						759.461	
40																						759.461	
20																						759.461	

Fig. 3.23 Analysis of the maximum stress combination for rider and passenger

R_1	54	50	45	40	35	30	25	20	15	10	5	0	-5	-10	-15	-20	-25	-30	-35	-40	-45	-50	-54
200	-	-	-	-	-	-	-	-	-	-	-	-	-	-	-	-	-	-	-	-	-	-	3,106
180	-	-	-	-	-	-	-	-	-	-	-	-	-	-	-	-	-	-	-	-	-	-	3,106
160	3,106	-	-	-	-	-	-	-	-	-	-	-	-	-	-	-	-	-	-	-	-	-	-
140	3,106	-	-	-	-	-	-	-	-	-	-	-	-	-	-	-	-	-	-	-	-	-	-
120	3,106	-	-	-	-	-	-	-	-	-	-	-	-	-	-	-	-	-	-	-	-	-	-
100	-	-	-	-	-	-	-	-	-	-	-	-	-	-	-	-	-	-	-	-	-	-	3,106
80	3,106	-	-	-	-	-	-	-	-	-	-	-	-	-	-	-	-	-	-	-	-	-	-
60	3,106	-	-	-	-	-	-	-	-	-	-	-	-	-	-	-	-	-	-	-	-	-	-
40	-	-	-	-	-	-	-	-	-	-	-	-	-	-	-	-	-	-	-	-	-	-	3,106
20	-	-	-	-	-	-	-	-	-	-	-	-	-	-	-	-	-	-	-	-	-	-	3,106

R_2	54	50	45	40	35	30	25	20	15	10	5	0	-5	-10	-15	-20	-25	-30	-35	-40	-45	-50	-54
200	-	-	-	-	-	-	-	-	-	-	-	7,080	-	-	-	-	-	-	-	-	-	-	-
180	-	-	-	-	-	-	-	-	-	-	-	7,857	-	-	-	-	-	-	-	-	-	-	-
160	-	-	-	-	-	-	-	-	-	-	-	8,829	-	-	-	-	-	-	-	-	-	-	-
140	-	-	-	-	-	-	-	-	-	-	-	10,079	-	-	-	-	-	-	-	-	-	-	-
120	-	-	-	-	-	-	-	-	-	-	-	11,746	-	-	-	-	-	-	-	-	-	-	-
100	-	-	-	-	-	-	-	-	-	-	-	14,071	-	-	-	-	-	-	-	-	-	-	-
80	-	-	-	-	-	16,266	-	-	-	-	-	-	-	-	-	-	-	-	-	-	-	-	-
60	-	-	-	-	-	16,266	-	-	-	-	-	-	-	-	-	-	-	-	-	-	-	-	-
40	-	-	-	-	-	-	-	-	-	-	-	-	-	-	-	-	#	-	-	-	-	-	-
20	-	-	-	-	-	16,266	-	-	-	-	-	-	-	-	-	-	-	-	-	-	-	-	-

M_f	54	50	45	40	35	30	25	20	15	10	5	0	-5	-10	-15	-20	-25	-30	-35	-40	-45	-50	-54
200	-	-	400.887	-	-	-	-	-	-	-	-	-	-	-	-	-	-	-	-	-	-	-	-
180	-	-	445.257	-	-	-	-	-	-	-	-	-	-	-	-	-	-	-	-	-	-	-	-
160	-	-	500.722	-	-	-	-	-	-	-	-	-	-	-	-	-	-	-	-	-	-	-	-
140	-	-	568.094	-	-	-	-	-	-	-	-	-	-	-	-	-	-	-	-	-	-	-	-
120	-	-	-	660.530	-	-	-	-	-	-	-	-	-	-	-	-	-	-	-	-	-	-	-
100	-	-	-	-	780.190	-	-	-	-	-	-	-	-	-	-	-	-	-	-	-	-	-	-
80	-	-	-	-	-	919.815	-	-	-	-	-	-	-	-	-	-	-	-	-	-	-	-	-
60	-	-	-	-	-	919.815	-	-	-	-	-	-	-	-	-	-	-	-	-	-	-	-	-
40	-	-	-	-	-	919.815	-	-	-	-	-	-	-	-	-	-	-	-	-	-	-	-	-
20	-	-	-	-	-	919.815	-	-	-	-	-	-	-	-	-	-	-	-	-	-	-	-	-

M_t	54	50	45	40	35	30	25	20	15	10	5	0	-5	-10	-15	-20	-25	-30	-35	-40	-45	-50	-54
200	-	-	-	-	-	-	-	-	-	-	-	-	-	-	-	-	-	-	-	-	-	472.591	-
180	-	-	-	-	-	-	-	-	-	-	-	-	-	-	-	-	-	-	-	-	-	472.591	-
160	-	-	-	-	-	-	-	-	-	-	-	-	-	-	-	-	-	-	-	-	-	472.591	-
140	-	-	-	-	-	-	-	-	-	-	-	-	-	-	-	-	-	-	-	-	-	472.591	-
120	-	-	-	-	-	-	-	-	-	-	-	-	-	-	-	-	-	-	-	-	-	472.591	-
100	-	-	-	-	-	-	-	-	-	-	-	-	-	-	-	-	-	-	-	-	-	472.591	-
80	-	-	-	-	-	-	-	-	-	-	-	-	-	-	-	-	-	-	-	-	-	472.591	-
60	-	-	-	-	-	-	-	-	-	-	-	-	-	-	-	-	-	-	-	-	-	472.591	-
40	-	-	-	-	-	-	-	-	-	-	-	-	-	-	-	-	-	-	-	-	-	472.591	-
20	-	-	-	-	-	-	-	-	-	-	-	-	-	-	-	-	-	-	-	-	-	472.591	-

Fig. 3.24 Analysis of the maximum stress combination for the sole rider

1. Roll angle $+50°$, speed 80 km/h (maximum R_1, mean R_2, M_f and M_t)
2. Roll angle $+40°$, speed 60 km/h (maximum R_2 and M_f)
3. Roll angle $-50°$, speed 80 km/h (maximum R_1 and M_t)

In the case of the sole rider (see table of Fig. 3.24 joint to data included from Figs. 3.15, 3.16, 3.17, and 3.18) the three combinations parameters are the following:

1. Roll angle $+54°$, all speeds (maximum R_1)
2. Roll angle $+30°$, speed 80 km/h (maximum R_2 and M_f)
3. Roll angle $-50°$, all speeds (maximum M_t)

Finally the stresses generated by the braking condition is independent from the roll angle (fixed at $0°$) and from the motorbike speed as clearly indicated by the data included from Figs. 3.11, 3.12, 3.13, and 3.14; the maximum stresses combination are, obviously given in the case of rider plus passenger.

Recalling the scheme of Fig. 2.13 of Chap. 2, the seven combinations of maximum stresses are, eventually, summarized in Tables 3.3, 3.4, 3.5, 3.6, 3.7, 3.8 and Figs. 3.25, 3.26, 3.27, 3.28, 3.29, 3.30 and 3.31. These loads combinations have to be used and imposed in the structural analysis which, normally, is carried out by means of a Finite Element Method (FEM) due to the high complexity of the shape and of the design of such a motorbike rear suspension.

Table 3.2 First condition of maximum stress for rider and passenger initial load

Description	Symbol	Value	Unit
Roll angle	α	+50	(°)
Motorbike speed	v	80	(km/h)
Bearing load	R_{t11}	3,829	(N)
Bearing load	R_{t12}	−68	(N)
Bearing load	R_{t21}	13,210	(N)
Bearing load	R_{t22}	5,115	(N)
Total axle load	R_{t1}	5,046	(N)
Total axle load	R_{t2}	9,381	(N)
Bending moment	M_f	604.864	(Nm)
Torque moment	M_t	−184.006	(Nm)
Spring load	$F_{rear} \times \cos(\theta_8 - \theta_1)$	6,534	(N)
Spring load	$F_{rear} \times \sin(\theta_8 - \theta_1)$	8,025	(N)
Connecting rod load	$F'_{rear} \times \cos(\theta_6 - \theta_1)$	7,240	(N)
Connecting rod load	$F'_{rear} \times \sin(\theta_6 - \theta_1)$	4,430	(N)

Table 3.3 Second condition of maximum stress for rider and passenger initial load

Description	Symbol	Value	Unit
Roll angle	α	+40	(°)
Motorbike speed	v	60	(km/h)
Bearing load	R_{t11}	5,838	(N)
Bearing load	R_{t12}	1,140	(N)
Bearing load	R_{t21}	22,206	(N)
Bearing load	R_{t22}	3,629	(N)
Total axle load	R_{t1}	4,769	(N)
Total axle load	R_{t2}	16,368	(N)
Bending moment	M_f	995.561	(Nm)
Torque moment	M_t	−88.372	(Nm)
Spring load	$F_{rear} \times \cos(\theta_8 - \theta_1)$	6,212	(N)
Spring load	$F_{rear} \times \sin(\theta_8 - \theta_1)$	7,499	(N)
Connecting rod load	$F'_{rear} \times \cos(\theta_6 - \theta_1)$	6,757	(N)
Connecting rod load	$F'_{rear} \times \sin(\theta_6 - \theta_1)$	4,160	(N)

Table 3.4 Third condition of maximum stress for rider and passenger initial load

Description	Symbol	Value	Unit
Roll angle	α	−50	(°)
Motorbike speed	v	80	(km/h)
Bearing load	R_{t11}	−2,191	(N)
Bearing load	R_{t12}	13,220	(N)
Bearing load	R_{t21}	7,191	(N)
Bearing load	R_{t22}	−8,173	(N)
Total axle load	R_{t1}	5,046	(N)
Total axle load	R_{t2}	9,381	(N)
Bending moment	M_f	177.507	(Nm)
Torque moment	M_t	759.461	(Nm)
Spring load	$F_{rear} \times \cos(\theta_8 - \theta_1)$	6,534	(N)
Spring load	$F_{rear} \times \sin(\theta_8 - \theta_1)$	8,025	(N)
Connecting rod load	$F'_{rear} \times \cos(\theta_6 - \theta_1)$	7,240	(N)
Connecting rod load	$F'_{rear} \times \sin(\theta_6 - \theta_1)$	4,430	(N)

Table 3.5 First condition of maximum stress for the sole rider initial load

Description	Symbol	Value	Unit
Roll angle	α	+54	(°)
Bearing load	R_{t11}	770	(N)
Bearing load	R_{t12}	−321	(N)
Bearing load	R_{t21}	2,598	(N)
Bearing load	R_{t22}	3,427	(N)
Total hub load	R_{t1}	3,106	(N)
Total hub load	R_{t2}	1,828	(N)
Bending moment	M_f	119.547	(Nm)
Torque moment	M_t	−133.048	(Nm)
Spring load	$F_{rear} \times \cos(\theta_8 - \theta_1)$	4,351	(N)
Spring load	$F_{rear} \times \sin(\theta_8 - \theta_1)$	5,061	(N)
Connecting rod load	$F'_{rear} \times \cos(\theta_6 - \theta_1)$	4,535	(N)
Connecting rod load	$F'_{rear} \times \sin(\theta_6 - \theta_1)$	2,794	(N)

Table 3.6 Second condition of maximum stress for the sole rider initial load

Description	Symbol	Value	Unit
Roll angle	α	+30	(°)
Motorbike speed	v	80	(km/h)
Bearing load	R_{t11}	4,822	(N)
Bearing load	R_{t12}	1,694	(N)
Bearing load	R_{t21}	21,088	(N)
Bearing load	R_{t22}	968	(N)
Total hub load	R_{t1}	2,662	(N)
Total hub load	R_{t2}	16,266	(N)
Bending moment	M_f	919.815	(Nm)
Torque moment	M_t	25.789	(Nm)
Spring load	$F_{rear} \times \cos(\theta_8 - \theta_1)$	3,686	(N)
Spring load	$F_{rear} \times \sin(\theta_8 - \theta_1)$	4,186	(N)
Connecting rod load	$F'_{rear} \times \cos(\theta_6 - \theta_1)$	3,734	(N)
Connecting rod load	$F'_{rear} \times \sin(\theta_6 - \theta_1)$	2,272	(N)

Table 3.7 Third condition of maximum stress for the sole rider initial load

Description	Symbol	Value	Unit
Roll angle	α	−50	(°)
Bearing load	R_{t11}	−1,658	(N)
Bearing load	R_{t12}	8,190	(N)
Bearing load	R_{t21}	4,344	(N)
Bearing load	R_{t22}	−5,123	(N)
Total hub load	R_{t1}	3,067	(N)
Total hub load	R_{t2}	6,003	(N)
Bending moment	M_f	95.343	(Nm)
Torque moment	M_t	472.591	(Nm)
Spring load	$F_{rear} \times \cos(\theta_8 - \theta_1)$	4,267	(N)
Spring load	$F_{rear} \times \sin(\theta_8 - \theta_1)$	4,947	(N)
Connecting rod load	$F'_{rear} \times \cos(\theta_6 - \theta_1)$	4,430	(N)
Connecting rod load	$F'_{rear} \times \sin(\theta_6 - \theta_1)$	2,727	(N)

Table 3.8 Maximum stresses for the braking phase and the rider and passenger initial load

Description	Symbol	Value	Unit
Bearing load	R_{t11}	10,187	(N)
Bearing load	R_{t12}	5,487	(N)
Bearing load	R_{t21}	1,641	(N)
Bearing load	R_{t22}	−66	(N)
Total axle load	R_{t1}	5,421	(N)
Total axle load	R_{t2}	−8,546	(N)
Bending moment	M_f	419.892	(Nm)
Torque moment	M_t	197.127	(Nm)
Spring load	$F_{rear} \times \cos(\theta_8 - \theta_1)$	2,872	(N)
Spring load	$F_{rear} \times \sin(\theta_8 - \theta_1)$	3,096	(N)
Connecting rod load	$F'_{rear} \times \cos(\theta_6 - \theta_1)$	2,749	(N)
Connecting rod load	$F'_{rear} \times \sin(\theta_6 - \theta_1)$	1,527	(N)
Pin load	$F_{pin} \times \sin\theta_{13}$	487	(N)
Pin load	$F_{pin} \times \cos\theta_{13}$	4,860	(N)

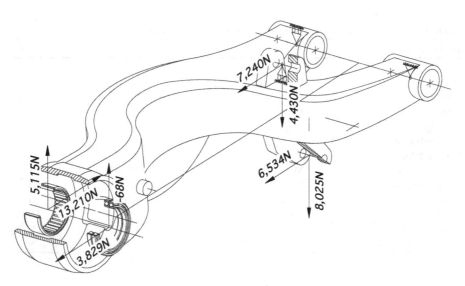

Fig. 3.25 First condition of maximum stress for rider and passenger initial load

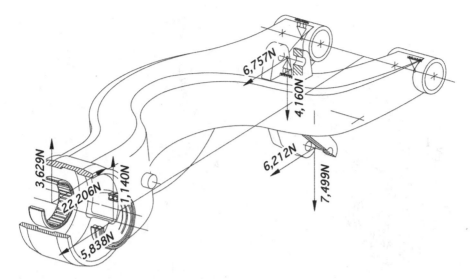

Fig. 3.26 Second condition of maximum stress for rider and passenger initial load

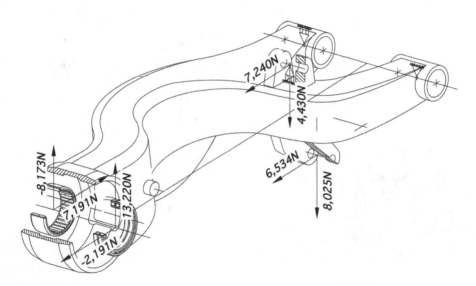

Fig. 3.27 Third condition of maximum stress for rider and passenger initial load

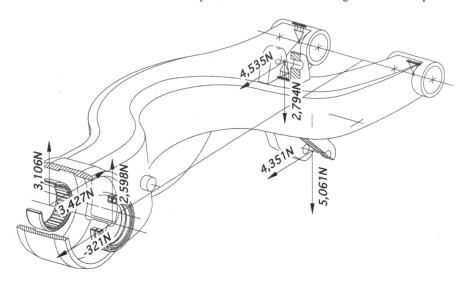

Fig. 3.28 First condition of maximum stress for the sole rider initial load

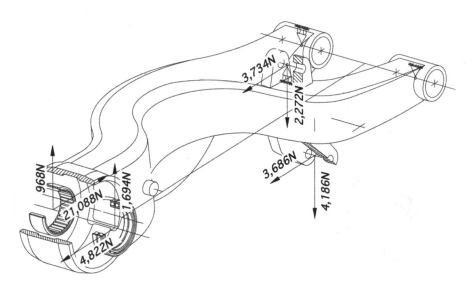

Fig. 3.29 Second condition of maximum stress for the sole rider initial load

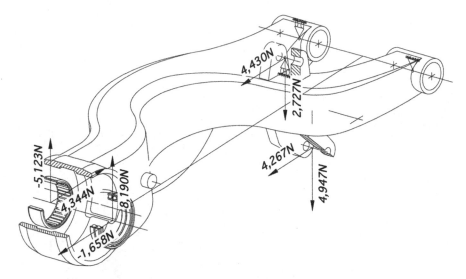

Fig. 3.30 Third condition of maximum stress for the sole rider initial load

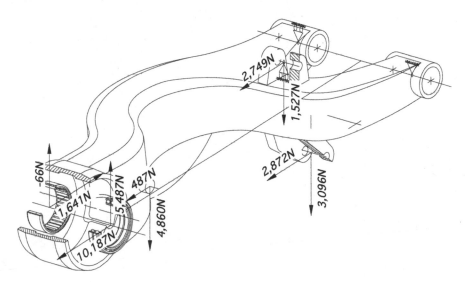

Fig. 3.31 Maximum stresses for the braking phase and the rider and passenger initial load

Chapter 4
The Front Suspension

Abstract In this chapter, the authors introduce the telehydraulic fork, which has become, across the years, the most common kind of front suspension for motorbikes. In chronological order, some noteworthy patents are presented, which help the reader following the main steps of the technical evolution which gave front motorbike suspensions their current shape. Then, the behaviour of some key structural elements of the fork is examined, under the assumption that the fork is subject to a bending moment acting on the front wheel mid-plane. Such a noteworthy loading condition is frequently encountered during the life cycle of a motorbike: consider, for instance, an emergency braking manoeuvre. Tests which simulate the effect of a hard braking on the fork are also part of the product validation programmes of the main motorbike producers. It is illustrated an analytical model useful for calculating the stress state of the fork legs under said loading condition. Such a model takes into account some architectural configurations as well as some characteristic geometrical parameters of the fork and of the motorbike. The model was validated referring to some production forks, both by finite element analyses and by experimental tests on the road. In the case of forks equipped with a single brake disc, the load unevenness between the legs during braking is analysed, and some strategies aimed at reducing such loading unbalance are suggested. The model presented herein could be helpful for designers who are developing new fork models, because it allows foreseeing critical issues due to legs dimensioning since the early phase of product development, when FEA techniques may be difficult to implement.

4.1 Historical Review

Since the late nineteenth century, engineers started to develop mechanisms aimed at improving the riding comfort of bicycles. Early solutions comprised one or more springs placed at some point of the bicycle frame, which enabled the wheels to rise and fall without communicating vibrations to the central part of the frame itself, the

D. Croccolo and M. De Agostinis, *Motorbike Suspensions*,
SpringerBriefs in Applied Sciences and Technology,
DOI: 10.1007/978-1-4471-5149-4_4, © The Author(s) 2013

Fig. 4.1 1901 bicycle with spring suspended handlebar, U.S. Patent No. 680,048, inventor Emil Koch

steering bar, the saddle, and therefore to the hand of the rider and to his body. An example of such devices can be appreciated in Fig. 4.1 which refers to a U.S. patent dating back to 1901 [1]. The mechanism shown in Fig. 4.1 may be regarded as a first form of front suspension. At that time, devices providing some sort of insulation of the saddle from the road irregularities were already available and covered by patents [2]. Unless the first motorbike appeared in 1869, we have to wait four decades to see the first example of front motorbike suspension, shaped as a fork [3]. It was introduced by the Scott Motorcycle Company of Shipley, West Yorkshire, England. Referring to Fig. 4.2, it can be seen a mechanism placed in front of a common bicycle fork, which carries the front wheel of the motorbike on the lower, U-shaped ends of its legs. The wheel is decoupled from the main frame of the bicycle by means of a couple of tubes sliding with respect to each other, sitting in the middle of the steering crown and attached on top of the steering pin. There are two coil springs inside the tubes, the outer spring is compressed when external loads are applied to the wheel, whereas the internal one is always loaded in tension and its only function is to check the recoil of the compression spring. Although original, this arrangement had little practical application and it is no longer in use. One of its major flaws, shared with its forerunners, is that a single, constant pitch, compression spring may hardly adapt to shocks characterized by an extremely wide amplitude range, like those encountered in a normal ride. In fact, if the spring was too stiff, it would not filter the road roughnesses when travelling over comparatively smooth roads. On the contrary, a highly flexible spring would be entirely compressed when clearing a severe bump or travelling over comparatively rough roads, thus transmitting undesired vibrations to the rider.

In 1914, Arthur Otto Feilbach, founder of Feilbach Motor Co. of Milwaukee, Wisconsin, obtained a patent in the United Kingdom for a motorbike fork comprising a spring preload adjustment mechanism [4]. Such a device can be seen in Fig. 4.3: by means of the nut 19, the spring preload can be adjusted by the rider to adapt to different tracks. Anyway, such a device would not allow to cure all the deficiencies listed above, because the spring setting is still constant during the whole compression and extension strokes.

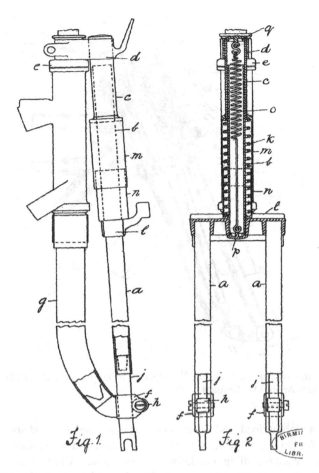

Fig. 4.2 1909 front fork with single coil spring for motorbike, G.B. Patent No. 7,845, inventor Alfred Angas Scott

In the early years of nineteenth century, a young apprentice worked at Feilbach's shop in Milwaukee: his name was William S. Harley, who surely leveraged some of the Feilbach's solutions for developing his own motorbikes. In fact, an invention patented by W. S. Harley in 1925 [5] was designed to overcome the aforementioned limitations. Referring to Fig. 4.4 it can be seen that this architecture is provided with two legs each side, and the front wheel is pivoting around a pin placed at the end of the rear leg. Figure 4.5 shows some sectional views of the suspension, where four different sets of springs can be identified. Each set is made up of two identical springs, and it is characterized by a certain number of coils, wire cross sectional area and spring index. These three characteristic parameters, along with the modulus of elasticity of the spring wire determine the spring stiffness. So the four sets of springs have different stiffnesses. The two sets of springs contained into the tubes have the lowest

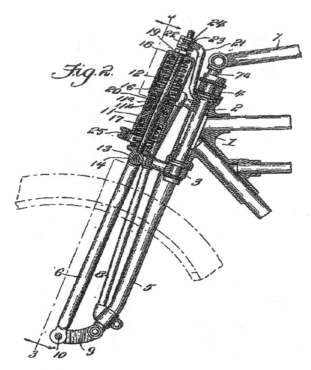

Fig. 4.3 1914 Improvements in Forks for Velocipedes, Motor Cycles and the like, G.B. Patent No.11,301 inventor Arthur Otto Feilbach

stiffness, but a high preload, and serve as compression and rebound springs (dashed line in Fig. 4.6). Owing to the high preload, they readily absorb minor shocks such as encountered when travelling on relatively smooth roads. Conversely, the two long external cushion springs, which sit on top of the lower fork clamp, are characterized by a higher stiffness and a much lower preload (dash-dot line in Fig. 4.6). Owing to the low preload, these springs do not become effective until a relatively severe bump is encountered. The springs of the shortest set are called bumper springs and come into action when very severe bumps are encountered, and the other sets of springs are almost fully compressed (thin solid line in Fig. 4.6). The overall elastic behaviour of the fork is represented by the solid thick line in Fig. 4.6. Although with some refinements, this architecture can be seen nowadays on some motorbikes produced by Harley Davidson Inc. Among others, the absence of any damping device, the relatively complex springs arrangement, the relatively high weight and a less than optimal serviceability may be cited as the main flaws of this architecture.

In 1934, Great Britain patent number 416,594 by Fisker and Nielsen [6] claims the invention of a telescopic fork with internal springs and easily detachable wheel axle for improved serviceability. This design was introduced on the 1934 Nimbus Type C motorbike. There are provided two springs and two telescopic legs (one spring per leg). Each leg ends with a U-shaped recess similar to that shown in Fig. 4.2.

Fig. 4.4 Side view of 1925 front fork for motorbike, U.S. Patent No. 1,527,133 inventor William S. Harley

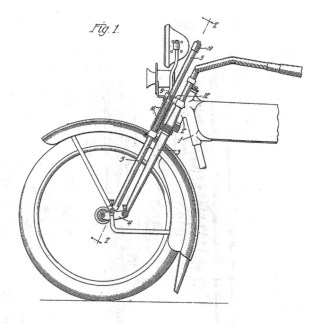

The wheel axle is secured to the legs by means of its threaded extremities. This improves serviceability, as the wheel can be easily removed by untightening two nuts on the wheel axle and pulling the wheel downwards. The solution is simple in design and relatively cheap to manufacture and assemble, as well as lighter than older designs which used external components and linkage systems. Its clean appearance makes it resemble modern motorcycle forks: nonetheless, it is expected that such a simpler spring arrangement can not ensure the same comfort performances as the invention of W. S. Harley shown previously.

A few years later, a milestone of suspension design was put forward by Rudolf Schleicher of Bayerische Motoren Werke (BMW) and covered by a deutsch patent in May 1939 [7]. This is the first example of telescopic fork with integrated hydraulic shock absorber, and the 1935 BMW R12 motorbike was already equipped with it. It consists of a telescopic fork inside which are added two tubes and a stem carrying a check valve at its lower end: these tubes are filled with a convenient amount of hydraulic fluid. Refferring to the notation in Fig. 4.7, the check valve 10 divides the lower fork tube 1 into a lower and an upper chamber. From the sectional view on the right side of the picture, it is possible to appreciate that a couple of sliding bearings support the movement of the telecopic tubes. During the compression stroke, as a consequence of the upward movement of the lower fork tube 3, of the shock absorber tube 1 and of the guide member 6, a depression is produced above the upper surface of the valve 10. Thus, the valve is drawn up and a free passage is created for the

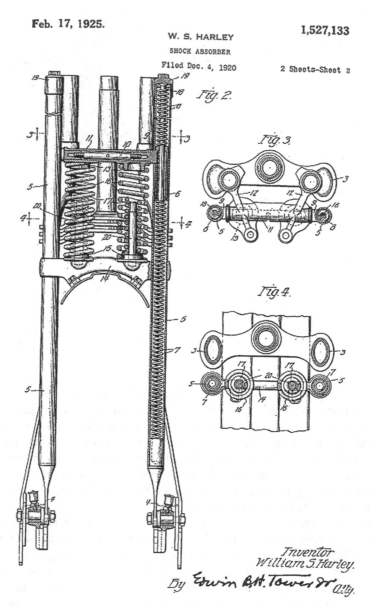

Feb. 17, 1925.

W. S. HARLEY

SHOCK ABSORBER

Filed Dec. 4, 1920

1,527,133

2 Sheets-Sheet 2

Fig. 4.5 Three section views of 1925 front fork for motorbike, U.S. Patent No. 1,527,133 inventor William S. Harley

hydraulic fluid flowing from the lower to the upper chamber. This means that during the compression stroke the fork moves without any damping effect. Conversely, during the rebound stroke, the downward movement of the three aforementioned

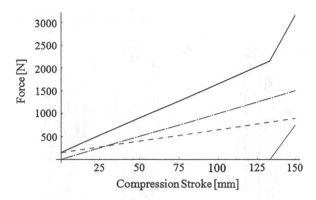

Fig. 4.6 Qualitative compression behaviour of a front fork like that of U.S. Patent No. 1,527,133

components makes the valve close, forcing the oil to flow through the small annular area between the stem 8 and the guide member 6, and eventually going back to the lower chamber by means of the comparatively large passages 15. Since the annular orifice section and the cross section of the valve are fixed, and the fluid kinematic viscosity is known, it is possible to calculate the pressure drop across the orifice, which is a function of the flow, and therefore of the sole rebound velocity. Such a valve arrangement is peculiar to motorbike forks, and, with slight differences, it can be found on some motorbikes still today. This solution had a few deficiencies too, as for example the dependency of the rebound damping level on the diametral gap between the guide member 6 and the stem 8. It must be remarked that the asymmetric behaviour between compression and rebound strokes is desirable, because relatively low damping in compression allows the wheel to follow the waviness of the road, ensuring a certain roadholding capability to the vehicle. At the same time, a stiffer damping on the rebound stroke helps cutting the oscillations of the vehicle body, thus improving comfort. Note that the spring lies outside the hydraulic damper and it is mounted with a certain preload between the lower fork clamp and the stanchion. The tube is retained to the fork clamps by means of a fixed ring under the lower clamp and a nut 9 tightened against the steering fork clamp. An evident drawback of this architecture is that the particular arrangement of the springs does not allow to have a long stroke of the tubular members. Therefore it can likely happen that such stroke would be insufficient when traveling over a rough asphalt surface or on offroad tracks.

Perhaps the solution proposed by three engineers of Dowty Equipment Ltd. in 1948 aimed at fixing some of those issues, by replacing the helical spring with a gas spring [8]. As it can be seen in Fig. 4.8 the legs are filled with oil up to the level marked by balloon number 33. An air chamber is formed at the top of each tube, and the two air chambers are in communication with each other through a duct realized inside the steering clamp. The air pressure can be adjusted to the desired level by means of the inflation valve 9. While the resilience of the fork is afforded by the compressed air, damping is due to the resistance to liquid flow created by the damping heads 16. Such damping heads shall have an established axial length, in order to provide a "frictional"

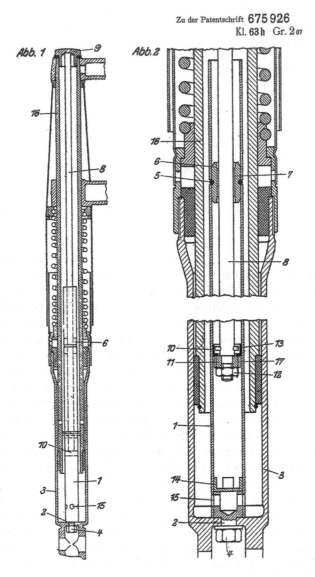

Fig. 4.7 Hydraulic shock absorber for motorcycle forks, DE Patentschrift 675,926, inventor Rudolf Schleicher

rather than an "orifice-like" damping. Since the fork has no springs, the inventors had to provide some means to arrest the sliding tubes at the stroke extremities. In fact, the damping head consists of a stack of Belleville washers: when, at the end of the extension stroke, said damping head comes into contact with the annular gland 13, the upward movement is arrested gradually, owing to the resilience of the washers. The same effect is obtained for the compression stroke by placing a rubber element

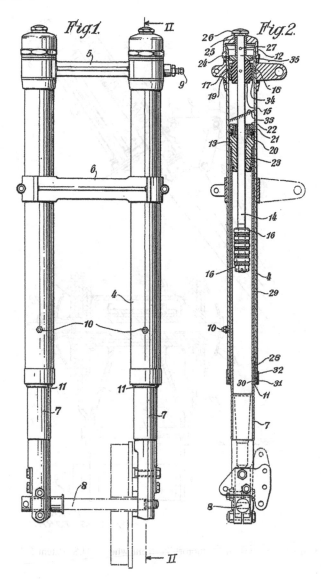

Fig. 4.8 An improved telescopic strut or shock absorber, GB Patent 597,036, inventors Peter Walter Burke, Richard Philip Wildey Morris and Arthur Adrian John Willit

15 beneath the upper cap 12. An attentive observer will surely notice that this one seems to be the first example of upside-down telescopic hydraulic fork in history. Upside-down forks have their legs arranged so that the outer tubes (or stanchions) are in the upper part of the fork and joined to the triple clamps. This solution provides a greater flexural stiffness to weight ratio, therefore, still today it is used in high end forks for sports motorbikes.

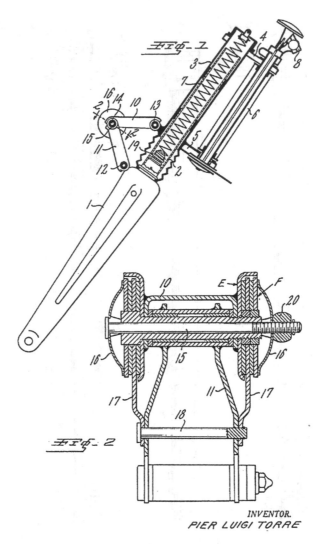

Fig. 4.9 Spring suspension system for motorbike front wheels, U.S. Patent 2,756,070, inventor Pier Luigi Torre

During the 1950s, various systems for replacing the elastic elements or the viscous damper were presented. A good example could be the invention by Pier Luigi Torre of 1956 [9]. Such architecture has a dry frictional damping element instead of the hydraulic one. There is provided a stack of frictional disks, like those present in a transmission clutch, held together by means of a threaded preloading mechanism which allows to adjust the damping effect to the rider's needs (Fig. 4.9). This solution had little fortune, because it can be no longer seen on any motorbike.

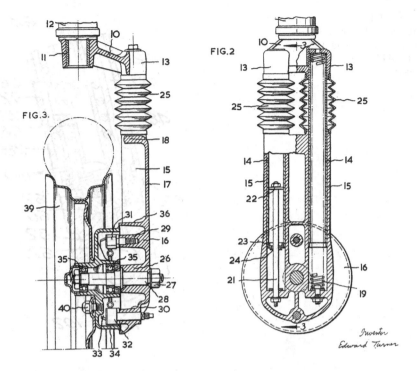

Fig. 4.10 Motorcycle front wheel suspension, U.S. Patent, 2,953,395 inventor Edward Turner

During the same years, some new designs for the front suspension of mopeds were presented, among which that by NSU Werke Aktiengesellschaft [10] and that by Edward Turner of Triumph Engineering Ltd. [11] are prominent. Both of them can not be defined forks, as they are basically single arm suspensions: two sectional views of the second one can be seen in Fig. 4.10. In spite of that, especially Turner's invention, originally applied to the 1958 Triumph Tigress (a scooter designed to have good performance and handling for the motorcycle enthusiast), deserves citation because it introduces a clever innovation. From the sectional view on the right of Fig. 4.10 it can be seen that there are actually two struts on the left side of the front wheel. One strut has a coil spring in it (the elastic element) while the other one has a viscous damper like those shown above. Therefore, instead of having hydraulic and elastic elements on both the struts, the two functions are separated and much less components are used. Being rigidly connected, the two elements still work in parallel as a mass-spring-damper system. Such a device is currently used for cheaper motorbike forks. It is worth observing that the tubes 14 are secured to the steering crown 13 by brazing or by interference fitting.

At the end of the sixties, some detail improvements were put forward with a new design by Joseph P. Roberts [12]. He developed a front fork comprising a hydraulic cushioning device working at extreme stroke positions and a lengthened helical spring.

Fig. 4.11 Telescoping, spring-loaded, hydraulically damped shock absorber, U.S. Patent, 3,447,797 inventor J. P. Roberts

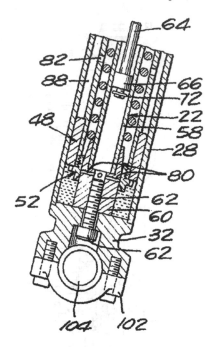

Referring to Fig. 4.11, which represents the check valve 66 and the stem 64 approaching the lower end of the stroke (the fork is almost fully compressed), it can be seen how the hydraulic fluid is trapped into a chamber created by the tube 82, the sliding bearing 48 and the outer tube 28. In the light of that, the oil can only flow to the low pressure chamber leaking between the sliding bearing and the outer tube. Such a small passage area determines a great pressure differential between the two chambers, and then a noticeable cushioning effect on the fork. The conical shape of the element number 60, makes the transition between low damping high damping conditions be smooth. Since the seventies, such a cushioning device has been found on almost every fork. As for the helical spring of increased length, it is conceived to allow for a greater stroke with respect to more classical designs. In fact, the spring lower end is sitting on the bottom of the outer tube, instead of on top of the tube 58, as it happened on previous designs. In the meanwhile, suspension designers began to experiment more sophisticated springing and damping devices, in order to fulfill the demand for increased roadholding capabilities coming from sports motorbikes enthusiasts and from competitions.

A good example [13] could be the fork patented in 1979 by the italian producer Arces S.r.l. (the acronym means Arturo Ceriani sospensioni). It is known that the great majority of motorbike forks are filled with hydraulic fluid up to a certain level inside the telescopic tubes. Above the hydraulic fluid there is air at ambient pressure. When a vehicle equipped with such a fork travels over a hard bump or clears a step, it can happen that, due to the rapid movement, some air bubbles, could be

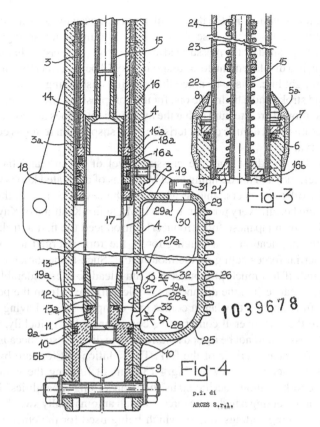

Fig. 4.12 Improved hydraulic fork, Brevetto per invenzione industriale, 1,039,678, licensee Arces S.r.l.

entrapped inside the hydraulic fluid. This results into a modified behaviour of the fluid, which generally determines a degradation of the suspension performances. Moreover, motorbike forks usually have two independent hydraulic circuits (one each leg), therefore, since the alteration of the fluid behaviour could be asymmetric for the two circuits, the resulting dynamic behaviour of the fork could be unbalanced and eventually dangerous for the rider. The invention by Arces consists in a fork having each leg provided with a sort of bladder accumulator, housed into a recess realized in the fork stanchion, as shown on the left side sectional view of Fig. 4.12.

In the same view, it can be seen that the hydraulic fluid is contained into the chamber 19 created between the stanchion 4, the tube 3 and the inner tube 15. As the tube dives into the stanchion (compression stroke), the oil flows through the check valve 27 and the flow regulator 32 to the accumulator chamber 26. Oil is pressurized by the action of the elastic bladder 29, inflated with compressed air via the check valve 31. During the rebound (or extension) stroke, the flow is inverted, then the fluid goes back to the chamber 15 by means of the check valve 28 and the flow regulator 33.

Both the inflation pressure and the flow regulators can be easily adjusted by the rider via some external knobs. In this case, the fork has no compression spring: the elastic function is accomplished by the pressurized air. Therefore, the elastic characteristic of the suspension, in a displacement-force diagram, would be described by an hyperbola instead of a straight line. It is to be noted that a spring cushioning device is provided for the rebound stroke and a hydraulic one for the compression stroke. The advantage of the aforementioned architecture is that the hydraulic fluid does not come in contact with air, therefore the hydraulic characteristic of the suspension remains constant for long time intervals.

The possibility to adjust both the pressure level of the air (and therefore the elastic characteristic) and the damping level by means of dedicated knobs was highly appreciated by off-road racers, among whom, in the seventies and in the eighties, such architecture became very popular. In 1981, Mitsuhiro Kashima of Kayaba Corp. Ltd. (a world renown japanese fork producer), introduced the first anti-dive system [14]. Under the application of a braking force, the fork is urged to compress in response to inertial forces acting on the motorcycle. Such a phenomenon is known as "diving", and, if too much pronounced, it may lead to a considerable increase of the braking distance (i.e. the distance a vehicle will travel from the point where its brakes are fully applied to when it comes to a complete stop). Diving is directly proportional to the fork overall compressive compliance, but unluckily, increasing the spring stiffness would not be a good solution to the diving issue, because it would affect the comfort characteristic of the fork. The solution put forward by Kashima and shown in the sectional view of Fig. 4.13 aims at limiting the natural diving tendency of the fork, without sacrificing its shock absorbing capabilities. The goal is achieved by implementing a hydraulic circuit which automatically switches between two different damping settings, one of which being used for the emergency brake

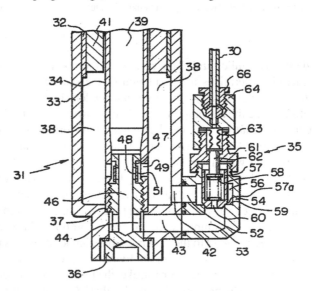

Fig. 4.13 Anti dive mechanism for motorbike forks, U.S. Patent, 4,295,658, inventor Mitsuhiro Kashima

condition and the other for the normal ride. The hydraulic circuit is made up of a first (38) and a second chamber (39). A normally closed check valve 57 divides the two chambers, and is forced to open by fluid pressure in the first chamber. As the fork is compressed, the hydraulic fluid flows from the first to the second chamber via said check valve. An actuator, responsive to the level of the braking force applied by the rider, increases the pressure required to open the check valve 57. This prevents the shock absorber from diving rapidly as a consequence of an emergency brake. As happens in any hydraulic circuit, shock absorbers comprise a great number of valves, working together to fulfill a precise function. In the case of vehicle suspensions, and especially those destined to motorbikes, a demanding task for a hydraulics engineer is managing to fit all the hydraulic functions in the smallest amount of space.

An interesting solution pointing towards that direction, was presented, among others, by Gert Neupert and Heinz Sydekum of Fichtel & Sachs AG, as it appears in an italian patent specification dating back to 1986 [15]. Such a device, shown in Fig. 4.14 consists in a rigid piston with two series of concentric axial passages. Each passage is free at one extremity and closed at the other by means of an elastic element which can be deformed in one direction only. The elastic members have different stiffnesses, therefore determining different pressure drops depending on

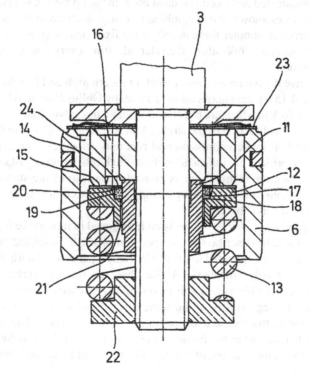

Fig. 4.14 Unidirectional/Flow regulator valve for shock absorber, Brevetto per invenzione industriale, 1,145,747, inventors Gert Neupert and Heinz Sydekum

the flow direction. The piston can be fixed to the shaft 3 of a twin tube damper or to the inner tube of a fork. In a few words, the valve arrangement works as a flow regulator valve in one direction and as a check valve in the other direction. Pistons like that of Fig. 4.14 can be seen in a wide variety of motorbike forks as well as rear shock absorbers, because they are compact, cheap to manufacture and characterized by a good reliability.

Nowadays such system is used, in more sophisticated forks, in combination with adjustable orifices which work in parallel with the elastic elements of the piston. Usually, the adjustable orifices are sized in order to saturate above a certain oil flow value (and therefore above a certain compression/rebound speed). When the orifices saturate, the oil is forced to flow through the passages on the piston, which determine a higher pressure drop. Doing that, it is possible to obtain a damping characteristic variable with relative the traveling speed of the suspension reciprocating members. A device of the aforementioned kind can be found in the patent by Adrianus H. I. Verkuylen entitled "Hydraulic shock damper assembly for use in vehicles" [16].

Todays motorbike forks come in various forms, depending on their mission: the majority of small motorbikes and mopeds are equipped with simple traditional forks, whose basic characteristics do not differ too much from the design of Fig. 4.11. On the contrary, modern sports motorbikes (including off-road vehicles) usually adopt much more complicated architectures, most often in the up side-down configuration, in which the devices shown above combine to bring about accurate damping characteristics. Moreover, simpler forks usually have fixed damping and elastic settings while high performance forks allow the rider adjusting every single aspect of their dynamic behaviour.

In order to give the reader an idea of what a modern high end fork looks like, it is reported in Fig. 4.15 a patent application of 1992, by Öhlins Racing AB [17]. That is an up side-down fork in which are used a couple of piston valves like that of Fig. 4.14, each working in one direction of the stroke. Moreover, in parallel with those valves, two orifices work at low compression and rebound velocities. The settings of the two low velocity orifices are independent from each other, and adjustable from two socket head screws, the rebound one is housed in the cap of the stanchion (which means it is accessible from the steering clamp), while the compression one is placed on the bottom of the axle bracket.

Although many improvements have been introduced in motorbike forks since the late 19th, it can still be described as a suspension member which includes a pair of telescoping tubes, one connected to the cycle frame by means of a triple clamp, and the other connected to the wheel axle. As the wheel receives a perturbation, the tubes are telescoped further within one another flowing an enclosed hydraulic fluid through one or more restricting orifices; at the same time, a coil spring is compressed. Oil and coil spring work together as a damper device, which serves to dampen shock and vibration effects received at the frame of the motorbike. In the next Section, topics concerning the structural design of a motorbike fork will be treated more in detail.

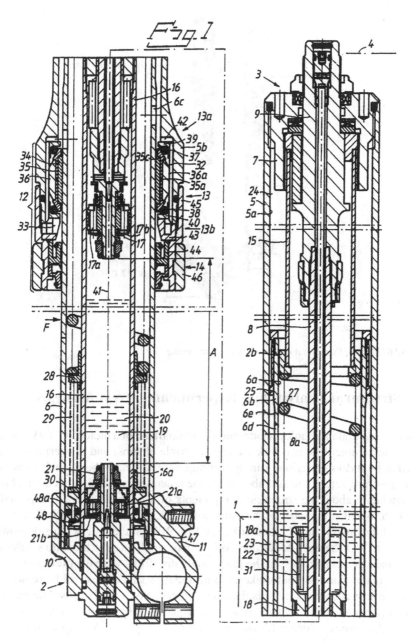

Fig. 4.15 Modern up side-down fork, PCT Patent application, WO 92/16770, inventors Kent Öhlins and Mats Larsson

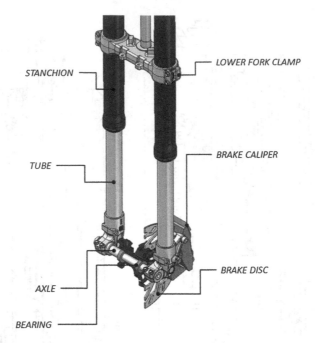

STANCHION

LOWER FORK CLAMP

TUBE

BRAKE CALIPER

AXLE

BRAKE DISC

BEARING

Fig. 4.16   Essential structural components of a motorbike fork

4.2 Structural Analysis and Experimental Stress Analysis

From a structural mechanics standpoint, the main frame of a motorbike fork consists of two telescopic legs, two fork clamps (or triple clamps) and a steering pin, put together by means of various joining techniques. The inner tube of each telescopic leg is generally referred to as "tube" while the other one is called "stanchion". Some nomenclature about the structural components of a motorbike fork is reported in Fig. 4.16. Whether the tubes or the stanchions are coupled with the fork clamps, the fork architecture is named "standard", see Fig. 4.17 on the left, or "upside-down", see Fig. 4.17 on the right. Motorbike forks can be subdivided further into single-disc and twindiscs architectures. In the past, single disc forks were the preferred choice for the great majority of motorbikes, but nowadays such devices will be reasonably found on low cost and small displacement motorbikes only, because the increase in performances and weight of sports motorbikes determined the success of twin discs architectures, which offer an increased braking power. A deep knowledge of the product lifecycle is the key for a correct design: for that reason, suspensions producers must carry out a number of road and bench tests in order to gather as much information as possible about their product. Several road tests carried out by the authors, led to defining the most severe loading conditions in terms of mechanical stress on the structural elements of the fork. In fact, the most part of field failures recorded on motorbike forks can be ascribed to bending loads on the legs, and happen

Fig. 4.17 Comparison between a standard fork (*left*) and an up side-down fork (*right*)

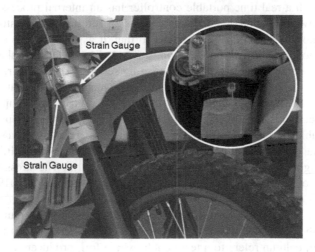

Fig. 4.18 Strain gauges installation

in the vicinity of the lower fork clamp [18]: such a kind of failure is often tied to fatigue phenomena and/or impact loads. Fork producers are usually challenged by their customers to provide design proposals for a new product: this process must be completed in a short time, because the product development phase shortened appreciably in recent years. During that early phase, no CAD geometries of the fork are yet available so that the preliminary structural design cannot leverage FEA analyses. Until a few years ago, manufacturers used to develop a new fork based on the characteristic dimensions of a previous model and then, by means of a trial and error process, assessing the validity of the new design by means of experimental tests run on pre-series prototypes. Therefore, the aim of the present work is to provide an analytical model which, based on a few parameters of the fork and of the bike, can help engineers to assess in advance the structural behaviour of the main components of the fork. In the past, such a methodology was applied by the authors at the University of Bologna, to provide mathematical models describing the behaviour of interference fitted connections between the steering pin and the fork clamps [19–21] and bolted connections between the fork clamps and the legs [22–24]. In order to assess the validity of the newly developed analytical model, it was compared with experimental data and with finite elements analyses (FEA). Extensive tests were carried out on an Enduro motorbike, equipped with the fork geometry shown in Fig. 4.16, which will be referred to as Fork 1. The Fork 1 is produced by Paioli Meccanica: it belongs to the up side-down type and is comes with a single brake disc. The fork was instrumented with four HBM 1-LY43-3/120 strain gauges placed underneath the joint between the lower fork clamp and the stanchions. The strain gauges were glued to the external surface of the stanchions, with the main grid axis aligned with the tube axis (Fig. 4.18). Each couple of strain gauges, half bridge configured [25], was connected to a NI 9237 C-DAQ module, plugged into a NI C-Rio 9014 real time portable controller. Such a real time portable controller has an internal processor: the data sampling software runs on that dedicated hardware and the data are stored into the internal memory of the controller. Thanks to that, it was possible to carry out the data acquisition without any interruption, downloading the data to the hard drive of a common notebook at the end of the riding session. The software, written by the authors in the Labview language, was used to manage all the data acquisition parameters: the sampling frequency was set at 100Hz. The bridge completion of the Wheatstone circuit is done by means of the C-DAQ module, which is provided with a reconfigurable internal circuit. Then, a professional rider performed a set of typical maneuvresmaneuvres, including the emergency braking on dry asphalt, the clearing of a step (height 150mm) and a jump from a height of 1m. The stress values were sampled for both the legs and the relevant plots are reported in Figs. 4.19, 4.20, 4.21 and 4.22.

The stress peaks, associated with each loading condition are summarized in Table 4.1 with subscript C (σ_{C_x}) for left leg and subscript D (σ_{D_x}) for right leg; the cam drum column refers to a test bench, which fork producers commonly use as a means for replicating the stresses produced on the fork by a certain road profile. In such a test bench, the whole fork group, along with the wheel rim and the tyre, is installed on a fixed support and then a rotating drum, to whose external circumference are applied obstacles of variable shape (cams), is put in contact with the tyre. As the drum starts, the fork is subjected to a series of repeated stresses, whose intensity is a function of the obstacles shape and whose frequency is a function of the angular velocity of the drum. Therefore, a run during which the fork

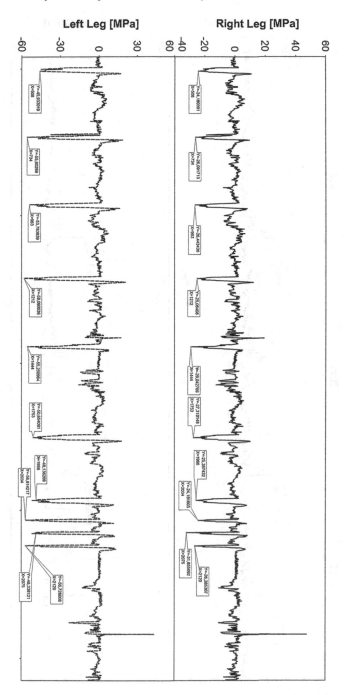

Fig. 4.19 Series of ten emergency brakings bending stress tracks for *right leg* (*solid line*) and *left leg* (*dashed line*)

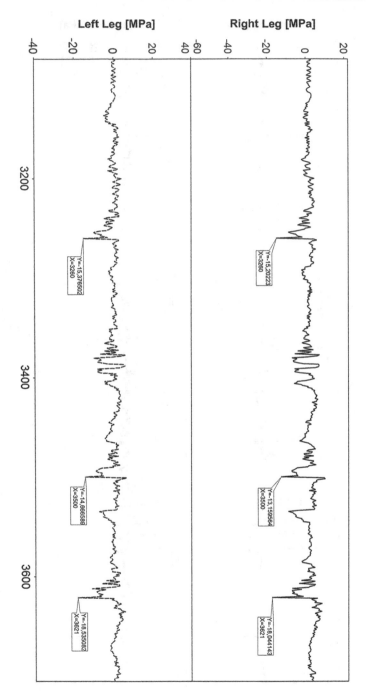

Fig. 4.20 Three step clearing maneuvres bending stress tracks for *right leg* (*solid line*) and *left leg* (*dashed line*)

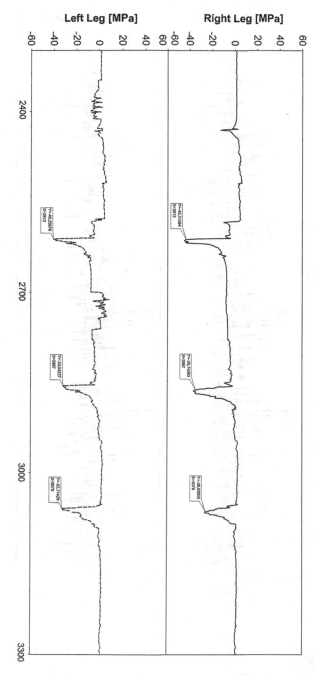

Fig. 4.21 Jump bending stress tracks for *right leg* (*solid line*) and *left leg* (*dashed line*)

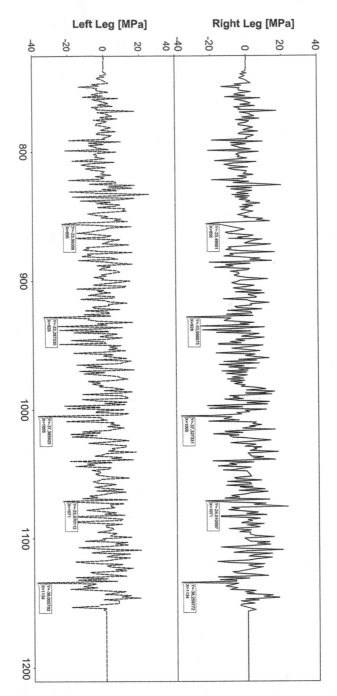

Fig. 4.22 Cam drum bending stress tracks for *right leg* (*solid line*) and *left leg* (*dashed line*)

Table 4.1 Average values for the maximum bending stresses

	Emergency braking	Step clearing	Jump	Cam drum
σ_{C_x} (MPa)	53	16	35	31
σ_{D_x} (MPa)	23	15	36	31

is subjected to several kinds of road irregularities is simulated. An attentive examination of Table 4.1 reveals that, only during the emergency braking, a considerable
stress unbalance exists between the two legs. Since the cross sections of the two
legs are identical, also the reaction moments in the vicinity of the joint between fork
clamp and stanchions must be different between the two legs. Such a behaviour is
peculiar to single disc forks and it is due to their asymmetric architecture: this implies
that the dimensioning of the stanchions cross sections cannot be optimized and that,
during the braking phase, the motorcycle tends to deviate from the correct trajectory because of the different deformation of the two legs. Since the average values
for the peak bending stresses of each loading condition are reported in Table 4.1,
it is possible to observe that the worst loading condition for the fork stanchions is
the emergency braking. Therefore, the analytical model proposed herein takes into
account the braking condition only.

4.3 Basics of Tyre Dynamics

In the light of the experimental analyses discussed in the previous section, there is
evidence that a hard braking manoeuvre, during which the rear tyre looses contact
with the ground, and the whole motorcycle weight is transferred to the front wheel,
determines the highest bending stress values on the fork legs. In order to build an
analytical model useful for the structural design of motorbike forks, such loading
condition must be deeply understood and described. For example, the adherence
characteristics of the tyre-road system, as well as the mass geometry of the motorbike,
play a significant role in defining the maximum bending loads due to a hard brake.
tyres are flexible elements which provide shock absorption while keeping the wheel in
close contact with the ground. tyres grip characteristics have a critical role in defining
the overall traction, braking and cornering performances that a motorbike can deliver.
Traction and braking forces arise during the ride, involving shear forces along the
contact area between the tyre and the ground. Such forces make the rubber fibres of the
tyre outer circumference compress along the tangential direction during the traction
phase and extend during the braking phase. Figure 4.23 shows a tyre travelling with
a velocity $V_0 = \omega \times r_0$, subject to a braking torque T_b which determines a braking
force F_b at the interface between the tyre and the ground. A vertical load N acts on
the tyre. Due to the longitudinal braking force F_b, the rubber fibres on the running
circumference elongate when passing through the tyre-ground contact segment AB

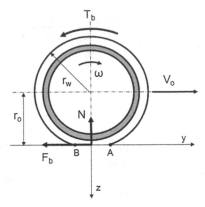

Fig. 4.23 Tyre subject to a brake

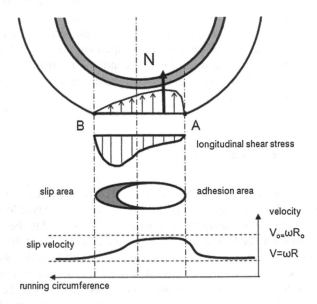

Fig. 4.24 Slip velocity during a brake

(Fig. 4.24). Hence, the circumferential velocity of a point fixed to the aforementioned fibres decreases as it travels from point A to point B: the circumferential velocity V of such a point is therefore lower than the travelling velocity V_0.

$$k = \frac{V - V_0}{V_0} \qquad (4.1)$$

Now, the longitudinal slip k defined in Eq. 4.1 takes positive values for traction and negative values for braking. Over the years, tyre manufacturers defined a variety of semi-empirical relationships expressing the longitudinal force F (and then the longitudinal friction coefficient μ') as a function of vertical load N and longitudinal slip values k. A widely known relationship of such nature is the so called "magic formula" due to H. B. Pacejka [26], a transcendental function generally expressed in the form:

$$F\langle k \rangle = D \times \sin\{C \times \arctan[B \times k - E \times (B \times k - \arctan(B \times k)]\} \qquad (4.2)$$

And the longitudinal friction coefficient μ':

$$\mu' = \frac{F}{N} \qquad (4.3)$$

Being F the longitudinal force and k the longitudinal slip. B, C, D, and E are input coefficients which depend on several "static" parameters (tied to the geometrical and chemical characteristics of the tyre) and on two "dynamic" parameters, namely the longitudinal slip k and the vertical load N. The "magic formula" 4.2 owes its name to the fact that there is no particular physical basis behind its mathematical form, but it fits a wide variety of tyre constructions and operating conditions. The equation can be plotted as shown in Fig. 4.25, where μ' is reported as a function of k. Looking at Fig. 4.25 it could also be noticed that the curve $\mu'\langle k \rangle$ is characterized by a peak, which is typically found for values of k around $k = \pm 0.15$.

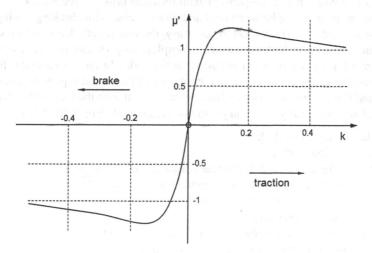

Fig. 4.25 Longitudinal friction coefficient μ as a function of the longitudinal slip k

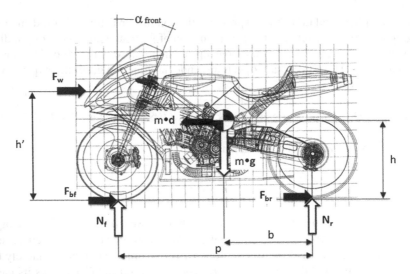

Fig. 4.26 Equilibrium of the motorbike during the emergency braking

4.4 The Analytical Model

Another element which influences the bending stress values on the fork legs is the amount of "diving" that the suspension exhibits under braking (see Sect. 4.1). In fact, the forces at the interface between road and tyre increase while braking, taking peak values at the end of the brake. On the contrary, the fork length decreases to become minimum when the vehicle comes to a complete stop (it can be easily observed by means of a linear transducer installed on the fork). In order to account for that behaviour, the analytical model shall comprehend a parameter representing the fork length and finite elements analyses shall be carried out with the fork both at the fully extended and at the fully retracted positions. Referring to Fig. 4.26, define:

- m: bike + rider mass [kg];
- p: bike wheelbase [mm];
- b: bike + rider centre of mass to rear wheel axis distance [mm];
- h: bike + rider centre of mass to ground distance [mm];
- $h\prime$: centre of the aerodynamic pressure [mm];
- F_w: aerodynamic force [N];
- F_{bf}, F_{br}: front wheel and rear wheel braking forces [N];
- N_f, N_r: weight on the front and on the rear axle [N].

When the motorbike stands on a flat surface, the vertical load acting on the rear wheel is:

$$N_r = m \times g \times \frac{p - b}{p} \qquad (4.4)$$

Define now the amount of load transfer from the rear axle to the front axle as:

$$\Delta N = m \times d \times \frac{h}{p} \qquad (4.5)$$

When brakes are applied, the vertical loads on each wheel are then:

$$N_{r\prime} = N_r - m \times d \times \frac{h}{p} \qquad (4.6)$$

$$N_{f\prime} = m \times g - N_{r\prime} \qquad (4.7)$$

The governing relationships for the in-plane equilibrium of the motorbike during braking are:

$$F_w + F_{bf} + F_{br} = m \times d$$
$$(4.8)$$
$$m \times d \times h = m \times g \times (p - b) + F_w \times h\prime$$

where:

$$F_{bf} = \mu\prime \times N_{f\prime}$$
$$(4.9)$$
$$F_{br} = \mu\prime \times N_{r\prime}$$

Here $\mu\prime$ is the longitudinal friction coefficient, as defined in Eq. 4.3. Therefore, a braking manoeuvre in which the rear tyre looses contact with the ground, and the entire weight of the motorbike is transferred to the front wheel is characterized by $N_{r\prime} = 0$ [29–31]. In that case, 4.8 transforms as follows:

$$F_{bf} = m \times d - F_w$$
$$(4.10)$$
$$m \times d \times h = m \times g \times \frac{(p - b)}{h} + F_w \times \frac{(h\prime - h)}{h}$$

Then, recalling 4.7 and 4.9 we obtain from 4.14 the maximum applicable braking force:

$$F_{bf\_max} = min \left[\mu\prime \times m \times g; \ m \times g \times \frac{(p - b)}{h} + F_w \times \frac{(h\prime - h)}{h} \right] \qquad (4.11)$$

Of which the first limiting condition is given by the adherence limit of the front tyre and the second by the capsizing limit of the motorbike. On a dry, clean asphalt surface, it is commonly assumed that standard sport-touring tyres for motorbikes can deliver a maximum longitudinal friction coefficient $\mu\prime \cong 1.5$ [30]. During a hard braking, a motorbike with comparatively long wheelbase and low center of mass (like, for example a "cruiser" type motorbike) would probably reach the skidding condition before capsizing, while a motorbike of the "hypersports" type would probably do the opposite. The maximum applicable braking force is a key parameter for fork

Fig. 4.27 Line body scheme
of a motorbike fork

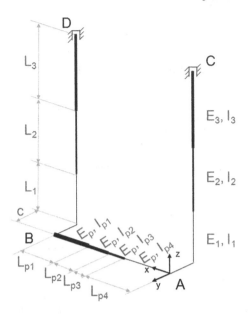

manufacturers, since it is strictly tied to the maximum bending loads that the fork
legs must withstand according to the product mission [27].

The analytical model was developed referring to the line body scheme shown in
Fig. 4.27; a portal frame [28] made up of three axisymmetric elements, the two pillars representing the legs and the transverse beam representing the axle. The relevant
Cartesian coordinate system is set in accordance with Fig. 4.27. The offset along the
y-axis, c, between the wheel axle and the leg axis is reported in Fig. 4.27 as well.
The pillars are constrained in C and D by means of two hinges, which allow the sole
rotation around the z-axis (as the two bearings, installed between the tube and the
stanchion, do in the actual component). The beam and the pillars have material and
inertia parameters variable along their axes, in order to achieve a reliable approx-
imation of the mechanical properties of the actual components. Each leg has been
subdivided into three segments: E_i and I_i being the Young's moduli and the x-axis
moments of inertia of the leg segments, respectively. In order to clarify what was
stated above, variable thickness lines represent the different properties of the portal
frame elements in Fig. 4.27. This complication is due to the fact that the leg is made
up of two elements, the tube and the stanchion, having variable materials and sections
along their axes (Fig. 4.28).

In order to achieve a correct positioning of the wheel hub and of the rolling
bearings, the axle has a variable diameter along its symmetry axis; the portal beam
was therefore subdivided into four sections with different moments of inertia (I_{pj},
J_{pj}). As shown in Fig. 4.28, the overall free length of the legs ($L = L1 + L2 + L3$)
is the distance, measured along the z-axis, from the wheel axle to the lower edge
of the lower fork clamp. When the brakes are activated, a load transfer from the

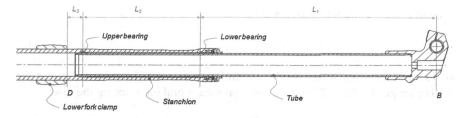

Fig. 4.28 Sectional view of a motorbike fork leg

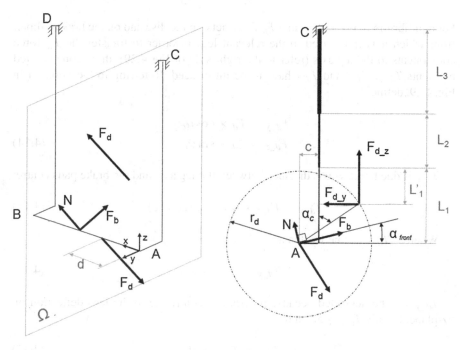

Fig. 4.29 Forces (*iso view*) and force components (*side view*) on the portal frame

rear wheel to the front wheel of the motorbike takes place. The amount of the load transfer depends on the center of mass position, on the motorbike wheelbase and on the deceleration value. Loads on the portal frame ar defined as for the iso view of Fig. 4.29: there is represented the portal frame, loaded out of its plane, with all the forces shown in their actual positions. In Fig. 4.29 is reported also a side view of the portal, showing the disc force components along with some useful dimensions. Referring to Fig. 4.29, r_d is the mean disc radius and α_c the angle between the z-axis and the brake pads centre (brake caliper angle); α_{front} is the caster angle (see also Fig. 4.26) and c is the offset between the axle and the leg axis. N and F_b forces are transmitted to the axle. In order to enforce the internal equilibrium, internal forces arise as well.

Define the braking torque on the disc:

$$T_b = F_{bf\_max} \times r_w \tag{4.12}$$

where r_w is the front wheel rolling circumference radius. In order to deliver the braking torque T_b, the braking pads must provide a braking force on the disc:

$$F_d = \frac{F_b}{r_d} = F_d \times \frac{r_w}{r_d} \tag{4.13}$$

For equilibrium reasons, the same F_d force acts on the disc and on the brake caliper, from which it is transferred to the relevant leg. In order to transfer the F_d force components to the leg axis (refer to the right side of Fig. 4.29), three concentrated moments T_{d_x}, T_{d_y} and T_{d_z} have to be introduced. Referring to the notation in Fig. 4.29, define:

$$F_{d\_y} = F_d \times \cos \alpha_c$$
$$F_{d\_z} = F_d \times \sin \alpha_c \tag{4.14}$$

T_{d_x} is due to the y-axis distance between the leg axis and the brake pads centre:

$$T_{d\_x} = F_{d\_z} \times (r_d \times \sin \alpha_c - c) \tag{4.15}$$

Whereas T_{d_y} can be written as:

$$T_{d\_y} = F_{d\_z} \times d \tag{4.16}$$

T_{d_y} will be neglected because it does not contribute to the legs deflection on yz-plane. Finally T_{d_z} is given by:

$$T_{d\_z} = F_{d\_y} \times d \tag{4.17}$$

The distance, measured along z-axis, between the F_d projection on y-axis, F_{d_y}, and the constraint C, is $L\prime = L_1\prime + L_2 + L_3$, where:

$$L_1\prime = L_1 - r_d \times \cos \alpha_c \tag{4.18}$$

In the following, force reactions will be represented by upper case letter R, while external moment reactions by upper case letter M. There are, therefore, six static equations and ten unknown force and moment reactions in C and D, leaving the system statically indeterminate. At the same time, it can be noticed that:

$$M_{C\_z} = M_{C\_z} = 0 \tag{4.19}$$

because the tube and the stanchion are free to rotate with respect to each other. R_{C_z} and R_{D_z} can be neglected because they are transmitted by the fork springs to the fork clamps, without affecting the flexural stress state of the legs. Moreover, the following external reaction components are null due to the loading condition:

$$R_{C\_x} = R_{D\_x} = 0$$

$$M_{C\_y} = M_{D\_y} = 0$$

(4.20)

The structure remains, however, statically indeterminate: the well-known method of consistent deformations [32] can be applied to solve it. The reaction moment values along the x-axis (M_{C_x} and M_{D_x}) are computed. The different intensities of such reaction moments testify that an uneven stress distribution between the two legs, introduced by the loading asymmetry, exists. The rotation angles around the x-axis of points A and B are defined by ϕ_A and ϕ_B respectively. The structure is uncoupled at points A and B and the consistent deformations equation is introduced:

$$\phi_A = \phi_B + \phi\prime$$

(4.21)

where $\phi\prime$ is the axle torsion angle between ends A and B. As formerly specified, a number of effects combine to bring about the overall M_{C_x} and M_{D_x} reaction moments; for instance F_b and N determine equal external reaction moments around the x-axis while F_d does not. Relying on the fact that the analysis is carried out in the elastic field, the aforementioned loads can be subdivided into four groups, applied to the structure separately, and eventually superimposed, in order to compute the overall M_{C_x} and M_{D_x} values. The first effect is due to the combined action of the disc force component F_{d_y} and of the concentrated moment T_{d_x}, both applied to the braking leg axis at $z = L_1 - L_1\prime$ (Fig. 4.30).

In order to provide the expressions of the reaction moments in C and D, it is useful to define the following two groups of parameters:

$$S_1 = \frac{L_3{}^2}{2 \times E_3 \times I_3} + \frac{(L_1 + L_2) \times L_3}{E_3 \times I_3} + \frac{L_2{}^2}{2 \times E_2 \times I_2} + \frac{L_1 \times L_2}{E_2 \times I_2} + \frac{L_1{}^2}{2 \times E_1 \times I_1}$$

$$S_2 = \frac{L_3}{E_3 \times I_3} + \frac{L_2}{E_2 \times I_2} + \frac{L_1}{E_1 \times I_1}$$

(4.22)

$$S_3 = \frac{L_p}{G_p \times J_p}$$

$$S_{11} = \frac{L_3{}^2}{2 \times E_3 \times I_3} + \frac{(L_1\prime + L_2) \times L_3}{E_3 \times I_3} + \frac{L_2{}^2}{2 \times E_2 \times I_2} + \frac{L_1\prime \times L_2}{E_2 \times I_2} + \frac{(L_1\prime{}^2)}{2 \times E_1 \times I_1}$$

(4.23)

$$S_{22} = \frac{L_3}{E_3 \times I_3} + \frac{L_2}{E_2 \times I_2} + \frac{L_1\prime}{E_1 \times I_1}$$

Then, the reaction moment due to the first effect in D can be expressed as follows:

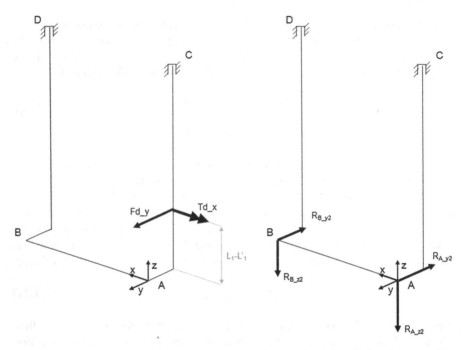

Fig. 4.30 Force and torque components on the portal frame – first effect (*left*) and second effect (*right*)

$$M_{D\_x1} = \frac{F_{d\_y} \times S_{11} + T_{d\_x} \times S_{22}}{2 \times S_2 + S_3} \ if \cos\alpha_c \geq 0$$

$$M_{D\_x1} = \frac{F_{d\_y} \times S_{11} + T_{d\_x} \times S_{22} - \left[\frac{F_{d\_y} \times (L_{1'}-L_1)^2}{2 \times E_1 \times I_1} - \frac{T_{d\_x} \times (L_{1'}-L_1)}{E_1 \times I_1} \right]}{2 \times S_2 + S_3} \ if \cos\alpha_c < 0$$

(4.24)

While the reaction moment in C is given by:

$$M_{C\_x1} = F_{d\_y} \times (L_{1'} + L_2 + L_3) + T_{d\_x} - M_{D\_x1}$$
(4.25)

F_d acts on the disc as well: then it is transmitted by the hub to the wheel axle and, finally, to the legs at points A and B (Fig. 4.30). The second effect is taken into account by means of the equations below, in which the principle of consistent deformations of Eq. 4.21 is applied again:

$$M_{D\_x2} = \left[\frac{(R_{A\_y2} - R_{B\_y2}) \times S_1 + (R_{A\_z2} - R_{B\_z2}) \times S_2 \times c}{2 \times S_2 + S_3} \right. $$
$$\left. + R_{B\_y2} \times (L_1 + L_2 + L_3) + R_{B\_z2} \times c \right] \quad (4.26)$$

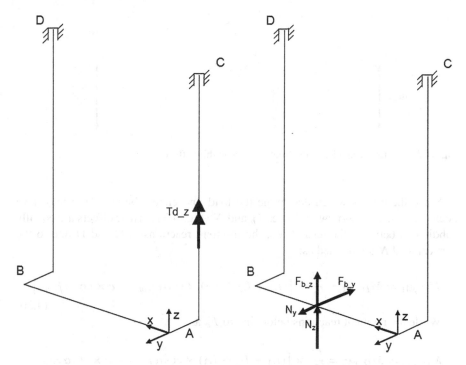

Fig. 4.31 Force and torque components on the portal frame − third effect (*left*) and fourth effect (*right*)

$$M_{C\_x2} = -\left[\frac{(R_{A\_y2} - R_{B\_y2}) \times S_1 + (R_{A\_z2} - R_{B\_z2}) \times S_2 \times c}{2 \times S_2 + S_3}\right.$$
$$\left. +R_{A\_y2} \times (L_1 + L_2 + L_3) + R_{A\_z2} \times c \right] \quad (4.27)$$

$R_{A\_y2}, R_{A\_z2}, R_{B\_y2}$ and R_{B_z2} forces can be computed by solving the equilibrium of the wheel axle (treated as a beam, simply supported at its ends A and B) subjected to F_d, applied at a distance d from A (Fig. 4.29). Looking at Fig. 4.31 T_{d_z} torque brings about the last contribution to the overall x-axis moment reactions in C and D, due to the disc force. Since the legs are allowed to rotate along the z-axis, they behave as simple supports for the wheel axle on the xy-plane. Hence, the concentrated torque can be considered as it was applied to the axle centre: the axle is treated, again, as a simply supported beam, now loaded by T_{d_z} (Fig. 4.32).

The moment reactions in C and D due to the third effect can be written as:

$$M_{D\_x3} = \left[\frac{(R_{A\_y3} - R_{B\_y3}) \times S_1}{2 \times S_2 + S_3}\right] + R_{B\_y3} \times (L_1 + L_2 + L_3) \quad (4.28)$$

$$M_{C\_x3} = -M_{D\_x3} \quad (4.29)$$

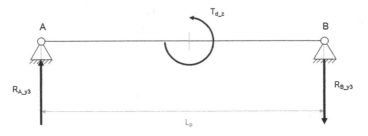

Fig. 4.32 T_{d_z} torque applied to the front wheel axle third effect

Since the effects which determine the load unevenness between the legs have been examined, focus now on forces F_b and N (Fig. 4.31). Their effects are equally subdivided between the legs. Then, the moment reactions in C and D due to the vertical load N are defined as:

$$M_{C\_x41} = M_{D\_x41} = \frac{N}{2} \times \left[(L_1 + L_2 + L_3) \times \sin \alpha_{front} - c \times \cos \alpha_{front}\right]$$
(4.30)

While the moment reactions belonging to F_b are:

$$M_{C\_x42} = M_{D\_x42} = \frac{F_b}{2} \times \left[(L_1 + L_2 + L_3) \times \cos \alpha_{front} - c \times \sin \alpha_{front}\right]$$
(4.31)

Then, the contribution of the centered loads is:

$$M_{C\_x4} = M_{C\_x41} + M_{C\_x42}$$
$$M_{D\_x4} = M_{D\_x41} + M_{D\_x42}$$
(4.32)

Superimposing the effects obtained in Eqs. 4.25, 4.27, 4.29 and 4.32, the overall reaction moments are expressed in the following Eq. 4.33:

$$M_{C\_x} = M_{C\_x1} + M_{C\_x2} + M_{C\_x3} + M_{C\_x4}$$
$$M_{D\_x} = M_{D\_x1} + M_{D\_x2} + M_{D\_x3} + M_{D\_x4}$$
(4.33)

Where the most important contribution to the overall moment reactions is given by the fourth effect.

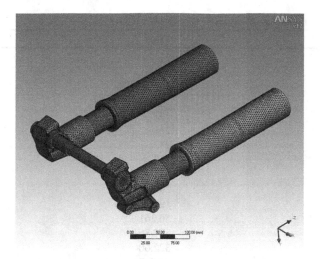

Fig. 4.33 Mesh for Fork 1 (shown retracted)

4.5 Numerical Validation of the Analytical Model

Once the formulae providing 4.33 are known, the stress values along the fork legs, or the moment reactions of the constraints C and D can be back-calculated. In order to validate the analytical model, a comparison with the results given by Finite Elements Analyses (FEA) of two different fork geometries is reported. The analyses were performed with the forks both in the extended and in the retracted configurations. If they were set up correctly, both the analytical model and the Finite Elements model, in the fully retracted configuration, should return the same z-axis normal stresses found on the outer tubes during the experimental tests.

The first geometry examined is shown in Fig. 4.16 (Fork 1). It is an upside- down fork, which means that the outer tubes are joined with the steering plates, while the inner tubes are joined to the axle by means of axle brackets. This fork carries a single brake caliper and it is conceived to suit small-medium "Enduro" motorbikes (125–250cc). All the Fork 1 data are reported in Table 4.2: these data are used as inputs for the analytical model. Table 4.3 reports the data of the motorbike in which Fork 1 was installed for experimental testing: these data are used to determine the force inputs for both the analytical and the numerical models.

FEA analyses were performed with the Ansys code, release 12.0: the model was meshed with tetrahedrons (Fig. 4.33) setting an element size of 3 mm for the tubes and the axle, and an element size of 4 mm for the stanchions, resulting into approximately 500,000 nodes. In order to implement finite elements analyses for both the extended and the retracted configurations, the pre-processing phase included developing two different CAD assemblies, one for each fork geometry. Then, F_d, N and F_b forces were applied to the structure conveniently, as shown in Fig. 4.34.

Table 4.2 Fork 1 parameters

Parameter	Value	Unit
α_{front}	26	(°)
α_c	85	(°)
c	34	(mm)
d	37	(mm)
$L_{1\_extended}$	397	(mm)
$L_{2\_extended}$	211	(mm)
$L_{3\_extended}$	24	(mm)
$L_{1\_retracted}$	137	(mm)
$L_{2\_retracted}$	211	(mm)
$L_{3\_retracted}$	24	(mm)
L_p	190	(mm)
I_1	56,261	(mm⁴)
I_2	378,861	(mm⁴)
I_3	322,600	(mm⁴)
J_p	19,769	(mm⁴)
E_1, E_p	206,000	(MPa)
E_3	71,000	(MPa)

Table 4.3 "Enduro" bike parameters

Parameter	Value	Unit
m	195	(kg)
p	1,475	(mm)
b	650	(mm)
h	650	(mm)
r_w	300	(mm)
r_d	117.5	(mm)
F_b	2,104	(N)

Two cylindrical supports, allowing the tubes to rotate around the z-axis, were applied, in C and D, to the stanchions (Fig. 4.34, left). F_d, the force acting on the axle, was subdivided into two components, which were each applied to a restricted surface of the axle corresponding to the contact area between the axle and the wheel bearing internal ring. Since all the geometries examined come from CAD assemblies, contacts between the parts needed to be defined. Such contacts were modelled as pure penalty, surface to surface, bonded. For the sake of realizing a model representing the overall structural behaviour of the fork, these contact setting is considered to be correct; moreover, it promotes convergence. In the light of that, the structure is considered as a single body and the results of the FEA can be compared directly with those provided by the analytical model. Such comparison is done for moment reactions in C and D, and reported in Table 4.4, which suggests a satisfactory accordance between the two methods.

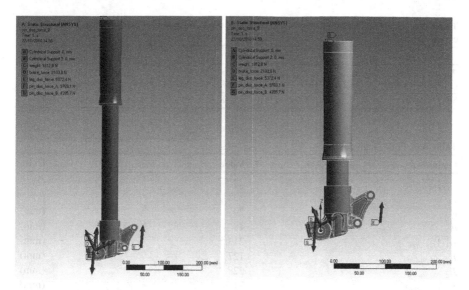

Fig. 4.34 Fork 1 FEA loads (**a**) extended, (**b**) retracted

Table 4.4 Moment reactions for Fork 1—analytical versus FEA

	FEA			Analytical				
	M_{C_x} (Nm)	M_{D_x} (Nm)	$M_{C\_x} + M_{D\_x}$ (Nm)	M_{C_x} (Nm)	ε (%)	M_{D_x} (Nm)	ε (%)	$M_{C\_x}+M_{D\_x}$ (Nm)
Extended	769 (64%)	436 (36%)	1205 (100%)	772 (64%)	+0.4	435 (36%)	−0.2	1,207 (100%)
Retracted	681 (73%)	253 (27%)	934 (100%)	683 (73%)	+0.3	250 (27%)	−1.2	933 (100%)

Table 4.4 shows that the braking leg always bears most of the total bending load; from 64 % (when the fork is extended) to 72 % (when the fork is retracted). In order to compare both the analytical model and FEA with experimental analyses (emergency braking column in Table 4.1), results in terms of normal stresses were evaluated and reported in Table 4.5. In the analytical model, such values can be computed easily 4.34, once the overall reaction moments for the retracted configuration are known:

$$\sigma_{C\_z} = \frac{M_{C\_x} \times D_{e\_3}}{2 \times I_3} \sigma_{D\_z} = \frac{M_{D\_x} \times D_{e\_3}}{2 \times I_3} \tag{4.34}$$

It was examined a second fork model (Fork2); it is a standard fork, to be installed on "Trials" motorbikes up to 250 cm^3. All the Fork2 data are reported in Table 4.6; these data are used as inputs for the analytical model. Table 4.7 reports the data of the motorbike for which Fork2 is designed.

The main differences between Fork1 and Fork2 can be summarized in: (i) Fork2 is a standard fork, in which the tubes are joined with the fork clamp and (ii) in the

Table 4.5 The z-axis normal stress values for Fork 1 (retracted)—analytical and FEA versus experimental

Experimental		FEA				Analytical			
σ_{C_x} (MPa)	σ_{D_x} (MPa)	σ_{C_x} (MPa)	ε (%)	σ_{D_x} (MPa)	ε (%)	σ_{C_x} (MPa)	ε (%)	σ_{D_x} (MPa)	ε (%)
53	23	57	+7.5	21	−8.7	59	+11.3	22	−4.3

Table 4.6 Fork2 parameters

Parameter	Value	Unit
α_{front}	22	(°)
α_c	338	(°)
c	29	(mm)
d	48	(mm)
$L_{1\_extended}$	213.5	(mm)
$L_{2\_extended}$	127	(mm)
$L_{3\_extended}$	228.7	(mm)
$L_{1\_retracted}$	31.5	(mm)
$L_{2\_retracted}$	309	(mm)
$L_{3\_retracted}$	46.7	(mm)
L_p	95	(mm)
I_1	135,157	(mm^4)
I_2	171,609	(mm^4)
I_3	36,452	(mm^4)
J_p	26,034	(mm^4)
E_1	71,000	(MPa)
E_3, E_p	206,000	(MPa)

case of Fork2, the brake caliper is placed on the front side of the leg (Fig. 4.35). Such a peculiar placement of the brake caliper si quite common in "Trials" motorbikes, because it offers a superior protection to the caliper when the vehicle travels over rugged roads (which often means rocky and steep mountain tracks). In this case, no experimental data are available, so the comparison is presented between analytical and FEA results only, both in terms of moment reactions and normal stresses. The FEA were carried out both on the extended and retracted configurations, with the same settings described above for Fork1: the mesh obtained is shown in Fig. 4.35.

The FEA were carried out both on the extended and retracted configurations, with the same settings described above for Fork1. Two cylindrical supports, allowing the tubes to rotate around the z-axis, were applied in C and D: since we are dealing now with a standard fork, the constraints shall be applied to the tubes and not to the stanchions. Results obtained for moment reactions in C and D are compared, in Table 4.8, with those given by the analytical model. Attentive examination of such results confirms the validity of the analytical model and that, again, the braking leg bears most of the total bending load; from 59 % (when the fork is extended) to 67 %

Table 4.7 "Trials" bike parameters

Parameter	Value	Unit
m	152	(kg)
p	1,305	(mm)
b	500	(mm)
h	750	(mm)
r_w	347	(mm)
r_d	82.5	(mm)
F_b	1,600	(N)

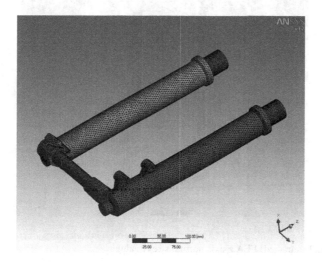

Fig. 4.35 Mesh for Fork2 (shown retracted)

Table 4.8 Moment reactions for Fork2—analytical versus FEA

	FEA			Analytical				
	M_{C_x} (Nm)	M_{D_x} (Nm)	$M_{C\_x}+M_{D\_x}$ (Nm)	M_{C_x} (Nm)	ε (%)	M_{D_x} (Nm)	ε (%)	$M_{C\_x}+M_{D\_x}$ (Nm)
Extended	598 (59 %)	421 (41 %)	1019 (100 %)	619 (60 %)	+3.5	406 (40 %)	−3.6	1,025 (100 %)
Retracted	569 (67 %)	283 (33 %)	852 (100 %)	602 (73 %)	+5.8	254 (30 %)	−10.2	856 (100 %)

(when the fork is retracted). Errors are slightly higher than those observed for Fork1, but they can be still considered acceptable ($\varepsilon\% \leq 10.2$).

Since no experimental data are available for Fork2, a comparison between the FEA and analytical results in terms of normal stresses is reported in Table 4.9. As for Fork1, such values can be computed by means of the analytical model, applying Eq. 4.34.

Table 4.9 The z-axis normal stress values for Fork2 (retracted)—analytical versus FEA

Experimental			Analytical		
σ_{C_x} (MPa)	σ_{D_x} (MPa)	σ_{C_x} (MPa)	ε (%)	σ_{D_x} (MPa)	ε (%)
311	150	314	+1.0	132	−12.0

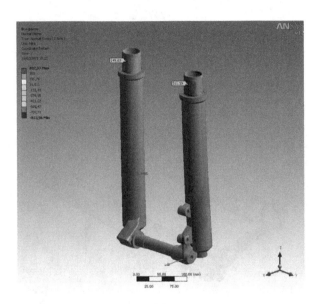

Fig. 4.36 Fork2 (retracted) FEA z-axis normal stress values

Looking at Table 4.9, the accuracy of the analytical model can still be considered satisfactory. Figure 4.36 is referred to one finite element analysis done on Fork2: the scale contours and the flags show normal stresses along the z-axis on both the legs, with the fork in the retracted configuration. As a general statement, the analytical model tends to slightly overestimate the loads on the braking leg while it underestimates the loads on the non braking leg. In this case, dealing with the z-axis normal stresses, values obtained in the vicinity of the constraints show much higher stresses with respect to Fork1: such behaviour is acceptable because the tubes are realized in 39NiCrMo3 steel ($S_Y \cong 680\,\text{MPa}$), while the stanchions are built in EN AW-6082-T6 aluminium alloy ($S_Y \cong 310\,\text{MPa}$).

4.6 Structural Optimization of Single Disc Forks

Tests made on different geometries returned similar results, showing differences between analytical and FEA stress distributions within a few percentage points.

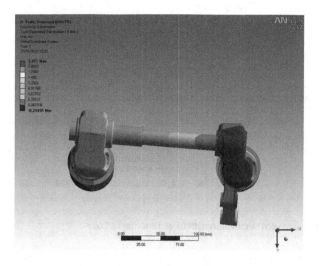

Fig. 4.37 Fork1 (*bottom view*) wheel axle undesired rotation around z-axis, due to loads unbalance

Where experimental analyses could be performed (Fork1), the results in terms of normal stress along the z-axis confirmed those given by the analytical model. Moreover, the newly developed analytical method allows fork designers to estimate the impact on the fork stress state of any design change upfront. The imbalanced loads during braking, characteristic of single disc forks, are an undesired effect, which yields different deformations of the legs and therefore a rotation of the axle around the z-axis (see Fig. 4.37). Such a rotation introduces an undesired steering action to the front wheel: hence, the motorbike tends to deviate from the correct trajectory while braking.

Thanks to the newly developed analytical model, overall moment reactions in C and D (or the relevant stress values) can be plotted as functions of a key design parameter of the fork. Therefore, design actions can be taken in order to minimize the load unbalance. Some screening tests were made in order to highlight which geometrical parameters affect the load unbalance significantly. In particular, the effects of axle offset c, disc offset d, moment of inertia of the axle I_p, and brake caliper angle α_c were examined. The moment of inertia of the axle I_p and brake caliper angle α_c were demonstrated to be the most influential factors. Moreover, the choice of the axle offset c and the disc offset d is often restricted by vehicle dynamics considerations and by characteristic dimensions of the brake caliper. In fact, it is very uncommon that a motorbike forks producer is also a braking systems producer. Then, brake calipers and discs are most often bought from original equipment manufacturers, who hold the design control over the characteristic dimensions of these products. In order to evaluate the influence of the moment of inertia of the axle I_p and brake caliper angle

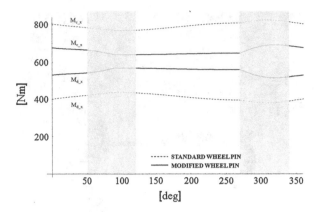

Fig. 4.38 Fork1 (*extended*) moment reactions as functions of α_c

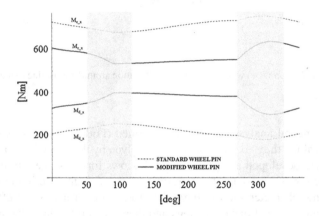

Fig. 4.39 Fork1 (*retracted*) moment reactions as functions of α_c

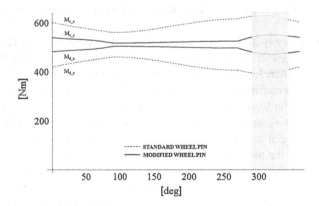

Fig. 4.40 Fork2 (*extended*) moment reactions as functions of α_c

α_c, the moment reactions for both the legs are plotted in Figs. 4.38 and 4.39 for Fork1 (extended and retracted, respectively) and in Figs. 4.40 and 4.41 for Fork2 (extended and retracted, respectively) as functions of the brake caliper angle α_c: dashed lines refer to a fork equipped with a standard axle whereas thick lines refer to the same fork equipped with a stiffer axle (moment of inertia ten times greater). Grey shaded areas in Figs. 4.38, 4.39, 4.40 and 4.41 indicate the ranges of acceptable values of α_c for each geometry. Such ranges are determined by manufacturability requirements as well as by product performance requirements. It should be appreciated that an even load distribution between the legs can never be reached, whatever the angular position of the brake caliper is (the curves never intersect). Moreover, the braking leg always bears higher loads than the non braking one. Nevertheless, the unbalance can be minimized by choosing $\alpha_c = 90°$ and increasing the stiffness of the axle. Such a behaviour can be understood by looking at Eq. 4.13, in which F_{d_y} becomes null for $\alpha_c = 90°$, and therefore T_{d_z} (Eq. 4.17) becomes null. In fact, the effect of T_{d_z} on the moment reactions is always equal in magnitude but with opposite directions on the two legs; when such an effect is null, the unbalance decreases. The effect of a stiffened axle is that of reducing the unbalance by transferring more load from the braking to the non braking leg.

The aforementioned strategies for reducing the unbalance have similar effects whether the fork be examined in the extended or in the retracted configuration; such an aspect can be appreciated by comparing, for example, Figs. 4.38 and 4.39. As shown by the same figures, the brake caliper placement in Fork1 was chosen quite well ($\alpha_c = 85°$; see Table 4.2). Conversely, for Fork2 ($\alpha_c = 338°$; see Table 4.6) a placement around the optimized value of $\alpha_c = 90$ deg was not permitted (see the shaded areas in Figs. 4.40 and 4.41), because of the product requirements said above. In fact, "Trials" motorbikes are subject to an extreme off-road use, and the brake caliper, as well as other components, must be placed as far as possible from the ground.

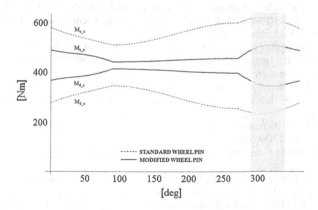

Fig. 4.41 Fork2 (*retracted*) moment reactions as functions of α_c

To summarize, from a structural standpoint, the emergency braking was found to be the most severe loading case for the bending stresses on the legs of motorbike forks. The novel analytical model here above described, allows calculations to be made of the stress field on the legs while braking, given a few geometric parameters of the motorbike and of the fork. The analytical model accounts for the unequal stress distribution on the two legs that arises from the geometrical asymmetry of single disc architectures. The model accuracy can be considered satisfactory, and was tested by comparison with the FEA and (if available) experimental stress analyses. The possibility of balancing the loads on the legs of single disc forks was examined as well. The brake caliper angle α_c and the axle stiffness are demonstrated to be the most influential parameters with respect to load distribution between the legs. Since an even load distribution between the legs cannot be achieved, some structural optimization strategies, targeted towards a significant reduction of the unbalance, are herein proposed. In the following chapter, it will be examined the stress state of the leg in the vicinity of the lower fork clamp, taking into account both the dynamic loads due to braking and the static loads due to the characteristics of the joint between the two parts.

References

1. Koch E (1901) Bicycle, U.S. Patent 680,048
2. Thompson ED (1898) Bicycle, U.S. Patent 598,186
3. Scott AA (1909) Improvements in or connected with the Front Forks of Motor-cycles, G.B. Patent 7,845
4. Feilbach AO (1914) Improvements in Forks for Velocipedes, motor cycles and the like, mG.B. Patent, 11,301
5. Harley WS (1925) Shock Absorber, U.S. Patent, 1,527,133
6. Nielsen AF, Fisker PA (1934) Improvements in and relating to handles and front fork for cycles, particularly for motor cycles, G.B. Patent, 416,594
7. Schleicher R Flüssigkeitsstoßdämpfer für. Kraftradgabeln, DE Patentschrift 675, 926, 1939
8. Burke PW, Morris RPW, AAJ Willitt (1948) Dowty Equipment Limited, An improved tele-scopic strut or shock absorber, G.B. Patent, 597,036
9. Torre PL (1956) Spring suspension system for motorbike front wheels, U.S. Patent, 2,756,070
10. Roder A (1958) Schwinghebel Federgabellagerung, insbesondere für das Vorderrad von Motor-rädern oder Motorrollern, DE Patentschrift, 1,043,844
11. Turner E (1960) Motorcycle front wheel suspension, U.S. Patent, 2,953,395
12. Roberts JP (1968) Telescoping, spring-loaded, hydraulically damped shock absorber, U.S. Patent, 3,447,797
13. Arces SRL (1979) Perfezionamento nelle sospensioni per motoveicoli, ITA Brevetto per inven-zione industriale, 1,039,678
14. Kashima M (1981) Front end shock absorbing apparatus for wheeled vehicle, U.S. Patent, 4,295,658
15. Neupert G, Sydekum H (1986) Valvola di trafilamento per ammortizzatori idraulici pneumatici o idropneumatici, ITA Brevetto per invenzione industriale, 1,145,747
16. Verkuylen AHI (1988) Hydraulic shock damper assembly for use in vehicles, U.S. Patent, 4,732,244
17. Öhlin K, Larsson M (1992) Device associated with a spring suspension system, PCT Patent application, WO 92/16770

18. Shelton H, Obie Sullivan J, Gall K (2004) Analysis of the fatigue failure of a mountain bike front shock. Eng. Fail Anal 11(3):375–386
19. Croccolo D, Cuppini R, Vincenzi N (2007) The design and optimization of fork-pin compression coupling in front motorbike suspensions. Finite Elem Anal Des 43(13):977–988
20. Croccolo D, Cuppini R, Vincenzi N (2008) Friction Coefficient Definition in Compression-fit Couplings Applying the DOE Method. Strain 44(2):170–179
21. Croccolo D, Vincenzi N (2009) A generalized theory for shaft-hub couplings. Proc Inst Mech Eng Part C J Mech Eng Sci 223(10):2231–2239
22. Croccolo D, Cuppini R, Vincenzi N (2009) Design improvement of clamped joints in front motorbike suspension based on FEM analysis. Finite Elem Anal Des 45(6–7):406–414
23. Croccolo D, De Agostinis M, Vincenzi N (2010) Recent improvements and design formulae applied to front motorbike suspensions. Eng Fail Anal 17(5):1173–1187
24. Croccolo D, De Agostinis M, Vincenzi N (2011) Failure analysis of bolted joints: Effect of friction coefficients in torque-preloading relationship. Eng Fail Anal 18(1):364–373
25. Doyle JF (2004) Modern experimental stress analysis. Wiley, Chichester
26. Pacejka HB (2002) Tire and vehicle dynamics. Butterworth-Heinemann, Oxford
27. Croccolo D, De Agostinis M, Vincenzi N (2012) An analytical approach to the structural design and optimization of motorbike forks. Proc Inst Mech Eng Part D J Automobile Eng 226(2):158–168
28. Belluzzi O, Zanichelli, Bologna, IT, Scienza delle costruzioni 1994
29. Cossalter V (2006) Lulu.com, Motorcycle, dynamics
30. Corno M, Savaresi SM, Tanelli M, Fabbri L (2008) On optimal motorcycle braking. Control Eng Pract 16(6):644–657
31. Cossalter V, Lot R, Maggio F (2004) On the stability of motorcycle during braking. In: Proceedings of the small engine technology conference and exhibition, Graz
32. Timoshenko S, Goodier JN (1951) Theory of elasticity. McGraw-Hill, New York